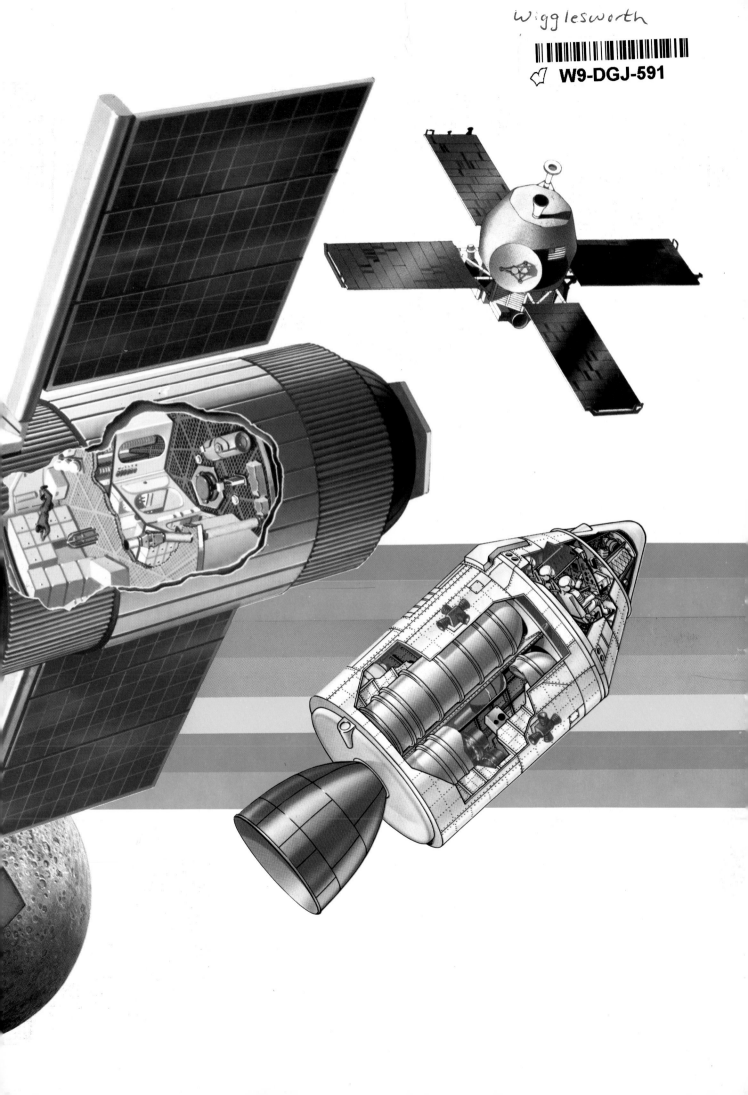

© Grisewood and Dempsey Ltd., 1989, 1982
All rights reserved. Printed and bound in Hong Kong
Published by Checkerboard Press, a division of Macmillan, Inc.
Library of Congress Catalog Card Number: 87-9397
ISBN 0-02-688538-7

CHECKERBOARD PRESS and colophon are trademarks of Macmillan, Inc.

CHECKERBOARD PRESS
Astronomy
Encyclopedia

CHECKERBOARD PRESS
New York

Contents

EDITORIAL
Frances M. Clapham
Ron Taylor
Cover: Denise Gardner

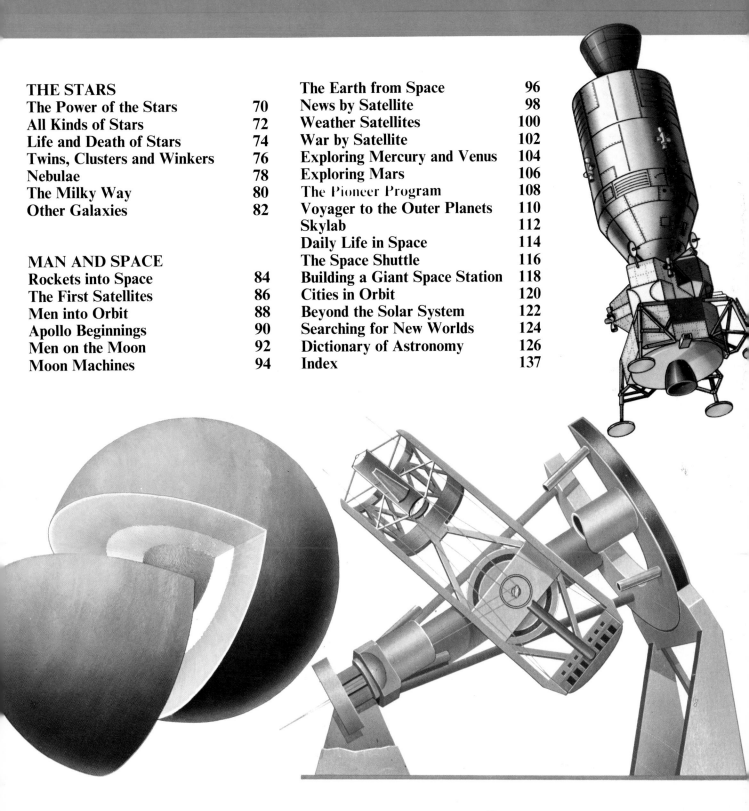

The Earth and Beyond

Our ancestors of the ancient civilizations knew little about the stars and planets. They gazed at the heavens and supposed that many gods and other supernatural beings lived there. Much later, during the scientific revolution of the 1600s, at the start of the Scientific Age, astronomers looked through their newly invented telescopes farther into the heavens. They saw, not supernatural beings, but what was surely just as exciting: entirely new and unsuspected details of the Moon and of Earth's fellow planets.

Nowadays, in the early years of the Space Age, astronauts can journey from the Earth's surface to spend long periods working in space, before returning safely to Earth. Before too long, it may be possible for anyone who wants a space vacation, and can afford one, to leave planet Earth for a while. The space shuttle, which made its maiden flight into space and back again as recently as April 1981, is the model for tourist spaceships of the near future, just as the first military jet aircraft of World War II were models for the airliners of today.

What it Feels Like in Space

In the early 1960s the first astronauts went into orbit around the Earth, more than 93 miles above its surface. They became the first human beings to live for any length of time in weightless conditions. This means that their bodies floated around freely, together with any other objects in their spacecraft that were not fixed down. In orbit, there is no force of gravity to weigh down objects. (Down here on Earth, the nearest most of us will ever come to a feeling of weightlessness is about halfway down from the high-diving board at the swimming pool.)

Scientists had known about weightlessness long before the first astronaut took off for space, but they were not at all sure what effects, bad or otherwise, it would have on an astronaut's fitness and health. We now know, after a number of astronauts have spent many months at a time in orbit, that their bodies *are* affected by the condition of weightlessness. For example, they may suffer at first from a sort of travel sickness known as space sickness. More seriously, after longer periods without gravity to pull their blood down into the lower parts of their bodies, the blood tends to collect or "pool" in the middle parts of their bodies. However, later on still, their bodies seem to adjust rather well and return to working more or less as normal. When the astronauts return to Earth after months in space, they certainly need to readjust to being heavy again. But so far, this has not stopped later astronauts spending even longer times in space.

Speeds in orbit are really tremendous, when compared with those here below. However, even if an astronaut leaves his spacecraft for a spacewalk, as the one in the picture opposite is doing, he will feel none of the familiar effects of high speed. This is because there is nothing nearby to give him the impression of speed as he whizzes past it. Neither is there any air to whistle in his ears, because he is orbiting above the Earth's atmosphere. And, because there is no air, no sounds come from anywhere outside the spacecraft. An orbiting astronaut lives in a strange, silent world, yet with the Earth always sunlit beneath him, and the stars always brilliant in the black sky, it is an amazingly vivid and impressive one.

Below: In November 1981 the space shuttle blasted off for orbit for the second time. As with the first flight, six months previously, the winged orbiter, Columbia, stayed a few days in space, then returned with its two-man crew safely to Earth base. In 1986 the space shuttle Challenger blew up after launch. No further shuttle launch took place until September 1988.

Left: An astronaut in orbit lives in a strange world where objects drift around without weight, free from the force of gravity. This applies to the astronaut himself, as the picture shows. In this case the astronaut is Jack Lousma, the Skylab 3 pilot. The photograph, taken in August 1973, shows him outside his spacecraft deploying the twin-pole solar shield which helped shade the orbital workshop. Reflected in his helmet visor is our Earth, the blue planet.

To get some idea of the sheer size of the Universe, it will help us to look at it in several stages, that get progressively larger. As the first stage, we can imagine a sphere of space within the Solar System that will easily hold the Earth. This sphere is about 15,535 miles across.

The next stage, a sphere 40,000 times larger, shows the Solar System as far as the orbit of the planet Jupiter. So this sphere is about 995 million miles wide, or more simply, rather less than 11 astronomical units wide.

How Big is the Universe?

"Universe" means "all the things that exist" and so when we talk about the Universe, we mean space plus its entire contents, to its very farthest extent. How far, then, does space extend? This sounds a direct enough question, yet its answer is not so simple.

For one thing, no telescope, however powerful, has yet allowed an astronomer to see to the farthest limit of the Universe. For another, the Universe is mostly empty space and it is difficult to imagine what could lie beyond it, when it came to an end – supposing, that is, that the Universe does have a limit, and does not stretch on forever.

Measuring Space
Despite these puzzling difficulties, it certainly is possible to give some kind of an answer to the question "How big is the Universe?" To measure anything, we first need a measuring rule or unit of some kind. To measure the lengths of small objects, we need a measuring rule graded in units such as inches or centimeters. For much longer distances here on Earth, we need to use units such as miles or kilometers. To measure cosmic distances – the sorts of distances encountered in outer space – we need to use much greater units still.

One of these units is the average distance of the Earth from the Sun: just under 92,957,102 miles. This is called the *astronomical unit* and is very useful for measuring distances inside the Solar System. Instead of saying that the distance of the planet Pluto from the Sun ranges from 2,796 million miles to 4,474 million miles, we can say much more simply that it ranges from 30 to 48 astronomical units. In this example, we are also measuring more simply the full extent of the Solar System, because Pluto is the most distant planet.

Beyond the Solar System, however, far greater distances are involved, which make even the astronomical unit too small to be useful for measurement. Even the nearest star beyond the Sun, Proxima Centauri, is 270,000 astronomical units distant, and more remote stars are impossible numbers of astronomical units away.

In a sphere 40,000 times as large again, the Sun would appear only as a lonely star in an empty space. This sphere is about four light-years across – the distance of the nearest star to the Sun.
The next sphere, 40,000 times yet larger, is about 160,000 light-years across and easily contains the whole of the Sun's galaxy. Our Solar System, as the pointer shows, is only a tiny speck in this galaxy, the Milky Way.

Our final magnification, again of 40,000 times, is a sphere with the enormous width of 600 million light-years. This includes about one third of the known Universe, and contains many nearer and more remote galaxies.

A still larger unit is needed for these farther off stars, and for other cosmic objects at still greater distances, such as galaxies beyond the Sun's own galaxy. The *light-year* is such a giant unit of distance. It is equal to the distance traveled by light in one year. Now, light travels at the enormous speed of 186,282 miles each second, so that the light-year is a correspondingly enormous unit. In round figures, one light-year equals nearly 6 trillion miles, or 63,000 astronomical units.

Using light-years, we can make measurements to the farthest limits yet reached by giant telescopes – although for the most remote objects, the figures do still get very large. A nearby star such as Proxima Centauri is a mere 4.3 light-years away from us, and the brightest star in the sky, Sirius the dog star, is only about twice as far, 8.7 light-years. The width of the Milky Way, the Sun's own galaxy, is about 1,000 light-years. The next galaxy to our own, known as Andromeda, has twice this width and is more than 2 million light-years away. The farthest-off galaxies and other celestial objects that astronomers can observe are 5,000 million or more light-years away from us. And still we have not reached "the end" of the Universe.

Even though the Universe is mostly empty space, it is so vast that it contains unimaginable numbers of stars. Our own Sun is a very average-sized yellow star, which is dwarfed by many giant or supergiant stars, as the diagram shows.

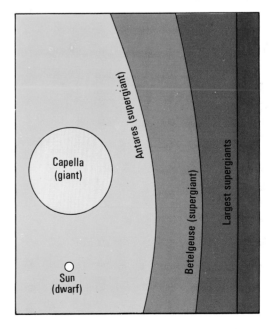

The Stars Above Us

Stargazing is a pleasure as old as the human race. On a clear, cloudless night, someone with normal eyesight will be able to see about 6,000 stars. Even the least powerful telescope or binoculars will show many times this number.

Just a few of the stars that we see, including some of the brightest, are not really stars at all, but planets. Brightest of all objects in the night sky, except for the Moon, is the planet Venus, which is also known as the Morning Star or Evening Star, depending on when it appears.

The giant planet Jupiter is very bright indeed, and the planets Mars and Saturn also shine brightly. The little planet Mercury can sometimes be seen low down in the sky, and the very distant planet Uranus is visible to some keen eyes.

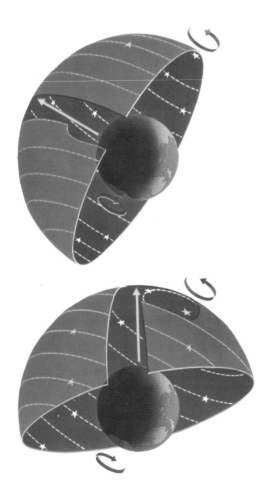

With only a little practice it becomes quite possible to tell apart stars and planets in the night sky. The four easily visible planets (excluding Mercury and Uranus) are fairly easy to remember. Venus and Jupiter are both brighter even than the brightest of real stars, the white-blue, twinkling Sirius. Mars can be recognized by its red color. Like the other planets, Saturn, although no brighter than a good many true stars, can be told apart by its wandering motion in the sky, which is very different from the more regular movement of the starry background. A time-exposure, or long-period, photograph of the night sky will show at once the difference between stars, which appear as simple streaks of light, and planets, which appear as more complicated, or wandering streaks.

Above left: Someone at the Equator looking up into the night sky will see the stars move across the sky in *straight* parallel lines. Below left: Someone looking up into the sky from farther north or south will see the stars move across the sky in *curved* parallel lines. In both cases, the star movement is really caused by the Earth spinning on its axis. The difference in star movement is explained by how near, or far, the observer is from the Poles around which the Earth spins. In the northern hemisphere, the Pole Star (yellow) will not move at all.

Starry Movements

Unlike people of long ago, we know that our own small world is very far from being the center of the Universe. But for the purpose of describing just where things are in the heavens, it is very convenient to pretend that the Earth is at the center of a vast starry globe or sphere. Astronomers call this the Celestial Sphere. Like a globe atlas, the Celestial Sphere has lines of latitude and longitude by which any particular point or area – say, a star or a gassy nebula – can be located exactly.

The Celestial Sphere can be used, among other things, to show the way that stars move in the sky. As the Earth spins on its axis, all the thousands of visible true stars will make the same movement. However, a stargazer will see the stars move rather differently according to whether he is nearer the Earth's Equator or the Earth's poles.

A second kind of star movement is less obvious, but photographs of the night sky, taken at intervals of days, weeks or months, will show it clearly. In this kind of movement, the positions of stars change because of the Earth's trip around the Sun, so that, after one year exactly, stars are back where they were before.

Above: Without a doubt, the first human beings gazed up into the night sky, as we still do now, and wondered at the multitude and mystery of the stars. Most of us can see about 6,000 stars without the aid of a telescope or binoculars – but these instruments will reveal many, many more.

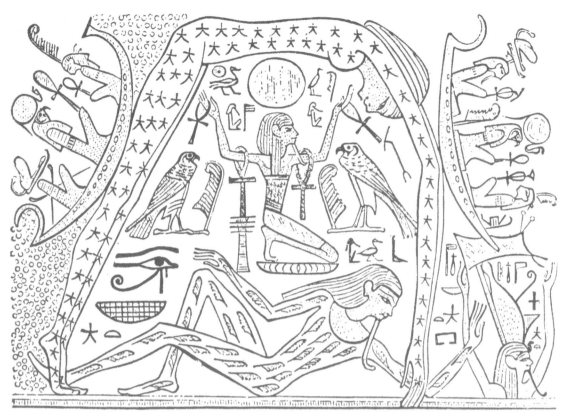

Early Stargazers

Modern astronomers are by no means the first to look with careful interest at the stars. We know that even the earliest civilizations paid close attention to what went on in the starry heavens. Very often, their scientific interest was mixed up with magic and religion, because they believed that the sky was one of the main homes of their gods. All the same, ancient stargazers were accurate observers, who left behind a great deal of useful information that has helped later astronomers.

Travel and Time

Although they had no telescopes or other scientific instruments, the ancient astronomers of Egypt and Mesopotamia made many useful discoveries and predictions more than 5,000 years ago. It was they who first noted that a few stars in the sky move very differently from all the others. These they named the wanderers, or planets. We now know that these are nearer to us than any true star, which explains their apparently erratic movements.

Ancient astronomers also named the first constellations, or patterns of different stars. This allowed any one bright star to be picked out more quickly from among the host of other stars. The identification of particular stars helped travelers to guide themselves safely by the starry heavens, and was especially useful to those who spent long periods out at sea, such as those greatest of ancient sailors, the Phoenicians.

Below: The ancient Mesopotamians were the first to name some of the constellations of stars. This stone tablet, 3,000 years old, shows the constellation of the Scorpion, as well as the Moon and the planet Venus.

14

Above: A Chimu disk showing a fertility goddess in the center, surrounded by eight panels showing farmers' activities at various times of the year. The Chimu are a rather mysterious people who lived in Peru in ancient times, in well-organized cities. From the look of this calendar, it seems as though they may also, like the ancient Egyptians, have studied the heavens for measuring the time of the year.

Times and Seasons

The first astronomers were also interested in measuring time. The cave dwellers of the Stone Ages had known that day and night are easily measured by the appearance and disappearance of the Sun. The early astronomers measured longer periods of the year with equal accuracy, by observing the different positions in the night sky of a number of constellations.

We now know that the movement of the constellations into different positions is really caused by the Earth's passage around the Sun. The first astronomers did not know this, but their astronomical observations still allowed them to tell the time of the year very accurately. By using the stars to set the times for planting, harvesting, and religious festivals, the early astronomers drew up the first calenders of important events throughout the year.

Eclipses

Still longer periods of time were measured accurately when the ancient astronomers noticed that eclipses also happened with a definite regularity. An eclipse takes place when one celestial or heavenly body hides another from view. The moon sometimes hides or eclipses the Sun, and a star may be eclipsed by a planet.

Eclipses usually happen over longer periods than a single year. Some only happen after an interval of many years. For this reason, and perhaps because eclipses were thought to have magical significance, ancient astronomers would often date an important event, such as a famous victory in battle or the crowning of a king, by a particular eclipse.

Such datings are of great interest to modern historians. Of more importance to modern science, however, is that this ancient interest in measuring and recording led to the development of mathematics and scientific astronomy. Gradually, the old astronomers became less concerned with superstitious beliefs about the heavens, and more interested in discovering the real nature of the Universe.

Right: The origin of Stonehenge, an ancient circle of stones on Salisbury Plain in southern England, was a great mystery for many years. It was thought to have been built by the Celtic Druids, perhaps for religious purposes – possibly even as a place where living people were sacrificed.

It now seems certain that the Druids had nothing at all to do with Stonehenge. It was built long before their time, between 4,000 and 3,000 years ago. Neither did Stonehenge have anything to do with sacrifices, although it may have served some religious purposes. Its main use appears to have been the measurement of the time of the seasons, by astronomical events such as the midsummer position of the Sun and the midwinter position of the Moon.

Above: The first scientific picture, or model, of the Universe was that of the ancient Greeks. This was perfected by the great astronomer Ptolemy of Alexandria. Ptolemy's Universe was a series of heavenly spheres, one inside the other. The stars were in the outermost sphere and the Earth was at the very center. The Sun went round the Earth, and so did the planets, each of which also had another motion, called an *epicycle*.

Eastern Astronomers

The first *scientific* astronomers were the ancient Greeks. This does not mean that they used many scientific instruments in their astronomy. It does not mean even that they discovered the true nature of the Universe. In fact, we accept very few Greek ideas about astronomy nowadays. But the ancient Greeks were the first to try to explain the Universe without the help of gods.

A Working Model of the Universe

About 2,500 years ago, ancient Greeks such as Pythagoras and Eudoxus invented a working model for the Universe. It was quite scientific – even though, as we now know, it was almost completely wrong. This model, or idea, is shown in the picture top left.

The ancient Greeks saw that the stars were farthest away, so they put them on the very outside. They thought that the stars were "fixed" – that they did not move. This of course was wrong, but understandable when you consider that the stars always seem to occupy the same places in the sky, year after year.

However, the Greeks' greatest mistake was to suppose that the Earth was at the center of the Universe, with the Sun and planets revolving around it. This was really a very old idea, even in ancient Greek times, since the Sun *seems* to go around the Earth, as it passes across the sky from sunrise to sunset.

The planets, though, have a different and more complicated movement, which is why they were named "planets," or wanderers. Rather later, ancient Greeks explained this by saying that the planets not only moved around the Earth in a great cycle, or orbit, but that they moved also in smaller cycles called epicycles. It was the combination of these two cyclic movements that made their wandering movement across the night sky.

Now that they had a working model of the Universe, the Greek astronomers could begin to examine its important details. About 140 BC, the Greek astronomer Hipparchus catalogued, or listed by name, no fewer than 850 stars. Two and a half centuries later another Greek, Ptolemy, listed 1,022 stars, and put the finishing touches to the Earth-centered model of the Universe. Wrong this model may have been, but it was very useful, compared with the old superstitions. Hipparchus and Ptolemy were the first great astronomers.

The ancient Chinese were the greatest inventors of their day. This Chinese tower, built about AD 1,090, was both a clock and an astronomical observatory. The clockwork was operated by water power, rather in the way of a water wheel. The clockwork in turn moved an armillary sphere on the roof of the tower. This sphere showed the changing positions of the Sun and planets.

Left: Hindu astronomers are best known for their improvement of the astronomical instruments used before the invention of the telescope. The picture shows two Hindu astronomers standing by their gnomon, or giant sundial. Like the much smaller sundials seen in some gardens today, this threw a shadow which showed the height of the Sun in the sky. At night, the astronomers used the gnomon to fix exactly the angle of certain stars so that they could make more accurate star maps.

After the Greeks

In Europe, people believed in the Greek model of the Earth-centered Universe for another 1,500 years; for after the Greeks, it was not in Europe that any real progress was made in astronomy. During the Dark Ages that kept Europe in ignorance, all further progress happened much farther east.

The Arabs kept Greek science alive by translating it into their own language, and keeping great libraries of scientific books. But Arab astronomers, like those in Europe, made few really new discoveries. They too held on to the Earth-centered model of the Universe.

Farther east still, Chinese astronomers owed much less to the Greeks and were more independent-minded. They bothered less about theories of the Universe, and concentrated more on making practical inventions for solving astronomical problems. Probably it was they who invented the armillary sphere, an iron or brass model of the Sun and planets in which the various movements of these heavenly bodies can be demonstrated. An armillary sphere is shown on the roof of the Chinese tower in the picture on page 16.

A more important Chinese invention was the magnetic compass, by which sailors could navigate without having to rely entirely on the stars. Still more important for the future of astronomy was the Chinese invention of glass lenses, which were to be used in the later invention of the telescope. This instrument was first made in Europe some centuries later, at the start of the modern scientific age.

Meanwhile, one other eastern civilization was to make useful progress in astronomy. Before the telescope, the angles and positions of stars in the sky had to be fixed using such instruments as the astrolabe and the gnomon. These instruments were greatly improved by the Hindu astronomers of India.

Below: In this picture, Arab astronomers of the Middle Ages are making various observations and calculations in astronomy and geography. At the top right of the picture, an astronomer is holding up an astrolabe, a circular instrument which was mainly used for navigation. Like the gnomon, the astrolabe measured the height, or angle, of the Sun in the sky. This allowed sailors to fix their positions at sea, and so make accurate maps. Several of the other astronomers are holding rulerlike instruments for measuring the angles of Sun, Moon, and stars.

Galileo's Breakthrough

As we have seen, the ancient Egyptians and Mesopotamians believed that the sky was the home of gods and demons. The ancient Greeks, whose ideas were more scientific, thought up a picture of the Universe with the Earth at the center and the Sun and planets revolving around it. This picture of the Universe was scientific because it was consistent. The Sun, Moon and stars moved in a regular way and were no longer upset by the behavior of demons or other spirits.

The Greek idea became the one accepted by astronomers for the next 1,500 years or so. But unfortunately for astronomy, the Earth-centered Universe became less and less of a scientific picture as the centuries went on. This was largely because the idea had now been adopted by the Christian Church as the way in which God had made the Universe. Because the Church said that the Earth-centered

Above: Nicolaus Copernicus, the 16th-century Polish astronomer who introduced the idea that the Earth revolves around the Sun, and not the other way around. This idea had first been suggested by the ancient Greek Aristarchus but had been ignored for nearly 2,000 years. Copernicus lived before the invention of the telescope, but he showed that the Sun-centered Solar System was the most logical one.

Below: Copernicus's Sun-centered Solar System is the Solar System we know today. It goes as far out from the Sun as Saturn, which was the last of the planets discovered in his day. A century after Copernicus, Galileo turned his newly invented telescope on to the Solar System and saw with his own eyes what Copernicus had only believed to be the case.

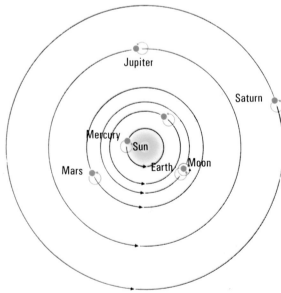

Left: Tycho Brahe, an astronomer who was born a few years after Copernicus died, also had no telescope with which to view the heavens. However, he did have the best astronomical instruments of his day. This one is for telling the heights or positions of stars in the sky. You can also see Brahe (three times) and his assistant (once) in the picture.

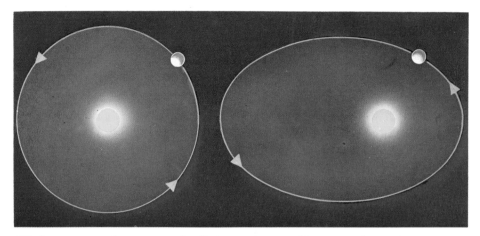

Below: Galileo Galilei was the first truly modern scientist. He founded modern physics, which later allowed Isaac Newton to discover the laws of motion and of gravity. He showed, using his telescope, that Copernicus had been right about the Sun-centered Solar System. The Church objected and made him a prisoner.

Universe was the true and religious picture, for a very long time no one even bothered to question it. Later still, in the 17th century AD, a Pope actually made it a mortal offense, or heresy, to deny that the Earth was at the center of the Universe.

But it was just at this time in history that scientists had succeeded in showing that the Earth-centered Universe was a mistaken idea. Early in the 16th century, a Polish astronomer named Nicolaus Copernicus wrote a large and detailed book with the title *Concerning the Revolutions of the Celestial Spheres.* In this book, he followed and improved on the ideas of the few ancient astronomers, notably Aristarchus of Samos, who had believed in a Sun-centered Universe. Copernicus showed in his book that the very complicated movements of the Earth's fellow planets in the night sky, which had puzzled astronomers for centuries, could be explained much more simply if the Earth and other planets were regarded as revolving about the Sun.

The Revolution in Astronomy

Copernicus's big book was revolutionary in more ways than one, because it upset completely the picture of the Universe preached by the powerful Church. But Copernicus was dead before his book came out, and not too much fuss was made about it at the time.

Things were very different about a century later, when the great scientist Galileo took up and championed Copernicus's Sun-centered Solar System. Lots had happened to advance astronomy in those hundred years. Tycho Brahe, observing the skies in Denmark, had provided a much more accurate and detailed picture of the stars and planets in the heavens – even though he still believed firmly in the Earth-centered Universe. Among Tycho Brahe's pupils was the best mathematician of his day. His name was Johannes Kepler, and unlike his teacher, he accepted Copernicus's ideas. Kepler turned his attention to those troublesome planets and showed that their orbits were not circles but *ellipses,* as shown in the picture above. This also contradicted the astronomy favored by the Church – but again, no fuss was made, probably because no churchman really understood what it was that Kepler had calculated.

At this point in history, Galileo Galilei first turned his new telescope on the skies. He saw, for the first time, Jupiter with its moons and Saturn with its rings move across the starlit heavens. This made him realize that the stars are infinitely remote, compared with the planets. What he saw also convinced him of the truth of Copernicus' Sun-centered Solar System. Galileo published his ideas and discoveries, which soon became world-famous. This was the point at which the Pope stepped in and banned all such ideas as heresy.

More New Ideas

Above: Isaac Newton was one of the greatest of all scientists. In this picture he is seen making one of his experiments with light, which led to his discovery and description of the spectrum – that white light is really made up of many colors. Among Newton's many other achievements was the discovery of the "Universal Law of Gravitation," by which he explained how the Sun and its planets and their moons move.

In the early 17th century, Galileo finally disproved the ancient idea of the Earth-centered Universe. The powerful Church stubbornly clung on to this idea for a while but finally had to give it up because so much evidence to the contrary was provided by other astronomers with their telescopes. This evidence pointed towards a Solar System in which Earth and its fellow planets went around the Sun. It also pointed to a Universe which was vastly bigger than our Solar System, and in which the stars were infinitely remote from us, compared with the other planets.

Galileo was not only the most revolutionary of astronomers, but was also one of the greatest of scientific experimenters. In one of his experiments he proved a scientific law which shows how fast and far a pendulum should swing. In another of his experiments, he rolled a wooden ball down a slope, then showed mathematically that the ball would always travel the same distance in the same time. But this is the same as proving how fast the ball gains speed in any particular time. In other words, by this experiment Galileo proved a scientific law of *acceleration*, or gain in speed.

At about the same time that Galileo was making his experiments, Johannes Kepler was making his careful calculations that showed that the orbits of planets and moons were not perfect circles, as had been supposed since the time of the ancient Greeks, but were "imperfect" ellipses.

Enter Isaac Newton

The first great scientific describer of the Universe lived in the generation which followed Galileo and Kepler, and founded his ideas on theirs. His name was Isaac Newton, and as most people know already, he discovered the law of gravity. This law is best known because it explains why all Earthly objects have weight – they are attracted by the Earth's force of gravity. But Newton's law of gravity applies to all matter – solid, liquid or gas – in the Universe. And it explains by how much the Sun attracts its planets, and the planets their moons.

Newton realized that Galileo's law of the pendulum is really an example of the law of gravity. The pendulum bob will swing just so far and so long as the Earth's gravity allows, before it is finally brought to a standstill. Newton also realized that Kepler's ellipses were just the kind of orbits that a moon would follow around its planet, or a planet around the Sun, when all these different worlds were held together in space by the force of gravity.

To complete his scientific description of the Universe, Newton needed to show just *how* all these movements occured – how fast, how far, how much force, and so on. Galileo had already discovered one universal law of motion, the law of acceleration. Newton completed this scientific. work by discovering the laws of motion for *all* kinds of movement here on Earth and in the rest of the Universe.

Below: Two hundred years after Newton, another scientist appeared who was to make equally great changes in our understanding of the Universe. At the youthful age of 26, Albert Einstein announced the first part of his "Theory of Relativity," in which he showed that space and time are not really different and separate, but can be regarded together as one and the same thing – *space-time*.

Enter Albert Einstein

Newton's description of the Universe, together with his other great mathematical achievements, ruled science for the next 200 years, until the beginnings of our own century. Even today, engineers use Newton's laws of motion and gravity, as for example when they work out how fast a rocket must be speeded up to go into orbit, or to escape from the Earth's gravity.

Only in our own century have Newton's laws come to seem less than completely universal. His laws still work perfectly for all motion at speeds up to that of a rocket leaving Earth. But for speeds much greater still, such as those nearing the speed of light, Newton's laws do not work accurately. This strange discovery was made by a young scientist named Albert Einstein, who went on to put forward his own revolutionary theory of the Universe – the *Theory of Relativity*. We look at this on the following page.

Right: A rocket blasts off on its journey through the atmosphere. Larger rockets can journey above the atmosphere into space, and perhaps into orbit around the Earth. How this can happen was first described by Isaac Newton, two and a half centuries before the first large rockets were invented.

Below: Newton's explanation went as follows. A stone thrown will follow a curved path, then fall to Earth because it is attracted, or pulled down, by the Earth's gravity (A). A shell fired from a gun will do likewise, but will follow a longer path (B) or (C). A shell fired with enough power (or, as we now know, a powerful enough rocket) will travel so far that when it begins to fall, it will fall with the same curvature as that of the Earth – in other words, it will fall into orbit around the Earth. Finally, if the shell or rocket has even greater speed than this, it will escape from the Earth's gravity altogether and be lost in space.

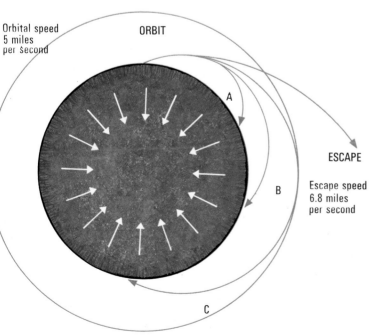

21

Gravity, Light, Motion and Time

The ideas of Albert Einstein were completely logical. They were also revolutionary because they completely changed our picture of the Universe and of space and time. Before Einstein, space and time had always been thought of as separate. Things exist both in space and in time, but these ways of existing were thought to be somehow different.

Einstein showed that space and time are not really separate but are two different aspects of *space-time*. That is, space and time are *relative* to one another – which is one of the reasons that Einstein's ideas are called the Theory of Relativity.

How can *we* understand this relativity of space and time? The easiest way is simply to look up into a clear night sky. What you see – thousands of stars – is a picture of both space and time, because the stars occupy thousands of different parts of space, and their light, traveling across the different distances, has taken different times to reach our eyes. The farther away a star, the longer its

Right: When you look into the night sky, you are looking into space-time. The farther away any object is in the heavens, the longer its light has taken to reach your eyes, and therefore, the longer ago you are seeing it. If you could see as far as the limits of the Universe, then you might also be seeing the beginnings of the Universe shown by the white question mark. Next come the farthest known galaxies, thousands of millions of light years away. Then we show our neighboring galaxy, Andromeda, 2 million light years away, and then the edge of the Milky Way, 20,000 light years away. The nearest star to us is only about four light years away, and the Sun is eight light minutes away from Earth.

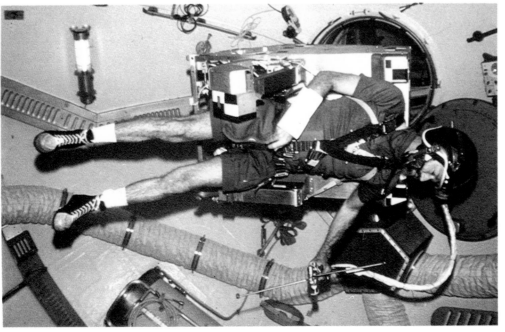

Some of the ideas in the Theory of Relativity are fairly hard to understand, but the relativity of weight is much easier to grasp. Generally speaking, weight is relative to gravity. The greater the pull of gravity, the heavier the weight of an object – although the *mass* of the object, its quantity of matter, does not change. The astronaut in this picture, in his spacecraft, is far beyond the pull of Earth's gravity; his body is weightless. The pull of gravity depends on the size of the body exerting it; on Earth, the astronaut would weigh some 143 pounds; on the smaller Moon, only 24 pounds.

Above: In the Universe, motion and non-motion are relative. You may be standing still, on Earth. But Earth, of course, is turning in space. Earth is also moving around the Sun. The Sun turns on its axis, and also journeys with its planets around our galaxy, the Milky Way. This galaxy itself, containing 100,000 million stars, spins in space and rushes away from other galaxies. So, wherever you "stood still" in the Universe, you would really be moving in some direction or other.

light has taken to reach us. So we are seeing each star at a different *time* as well as in a different *place* – and we can never see them any differently.

Einstein's theory says that because space and time are relative, a change in the one will produce a change in the other. In the picture bottom right, a rocket is shown speeding up in space. The Theory of Relativity says that its time, shown by a clock on board, will *slow down*. This seems an amazing result, because we ourselves do not experience such changes; they will become obvious only at speeds nearing the speed of light, or 186,282 miles per second. No human being has yet traveled at anything like such a speed.

The only things that do travel at such speeds are light itself and other forms of waves related to light, and subatomic particles such as electrons and protons. These subatomic particles can, for example, be boosted up to speeds near that of light, in giant machines called particle accelerators. Does their time really change, by slowing down? It is difficult to measure the time on a subatomic particle, because you cannot put a clock on it! But the Theory of Relativity says that mass or weight, and speed, are also relative. As an object speeds up, it gets heavier – which is also against all our own experience. Now it *is* possible for a scientist to measure the mass or weight of a subatomic particle, as this speeds up in an accelerator – and sure enough, when its speed gets near that of light, its mass shows a definite increase. Some other aspects of relativity are shown by the pictures.

Below: Einstein's Theory of Relativity says that time and speed are relative. The faster the speed of an object, the slower is its passage through time. But this is noticeable only with objects traveling very fast indeed. The picture shows the example of a rocket, with a clock on board. This takes off from Earth at 3 o'clock and speeds up very rapidly to about 139,800 miles per second – which is about three-quarters the speed of light. As the rocket reaches this huge speed, its clock will be showing 5 o'clock. But time has slowed down on the rocket, and the clocks back on Earth will now be showing 6 o'clock.

Space-Time Mysteries

Outer space has always been something of a mystery to people, ever since the earliest astronomers gazed into the night sky and wondered what was really there. To modern astronomers, with all their special scientific equipment, space is not less, but even more, of a mystery. This is because each new discovery made about space and time is more astounding than the last. We talk about space *and* time, because they are always linked together. If you look deep into the starlit night sky, you are also looking back into time. The farther away a star is, the longer its light has taken to reach your eyes. With a powerful telescope, you would see galaxies and other giant objects that are thousands of millions of light years away. So you would be seeing them as they were thousands of millions of years ago. Perhaps, if you had a telescope powerful enough (it has not yet been built) you could see back in time to the beginnings of the Universe – to the time of the Big Bang itself.

No one has yet seen this far, but the big telescopes of today have revealed mysteries enough, such as pulsars, quasars, and possibly the strangest "objects" yet – black holes.

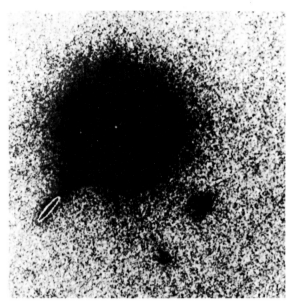

Above: This is a *negative* photograph, taken through a very powerful telescope, of one of the mysterious objects called quasars. These resemble stars (quasar means "like a star") but they give out far, far more energy. In fact, a single quasar gives out far more energy than a whole *galaxy* containing hundreds of thousands of millions of stars! Some astronomers think that quasars are connected with the even stranger black holes.

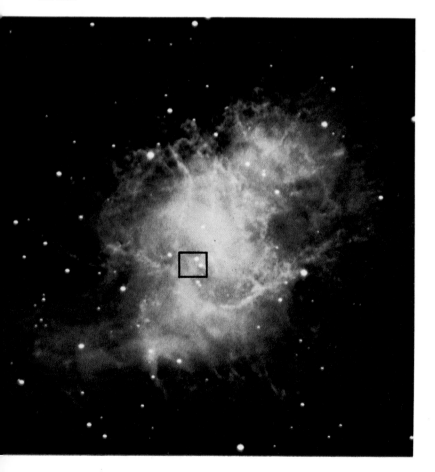

Left: The Crab nebula in the constellation Taurus became famous as early as July 4, 1054. It was on this date that astronomers first noticed it – because it then formed in the night sky as a clearly visible patch of light. This sudden information tells us that the Crab nebula is almost certainly the result of a giant star explosion, or supernova. Then, on February 24, 1987, another supernova was seen to explode in the Large Magellanic Cloud, a nearby galaxy.

Below: At the center of the Crab nebula (shown by the small square in the picture on the left) is a mysterious starlike object left over from the supernova. This is called a pulsar, from the fast pulsing with which it gives out energy. In the one picture the Crab pulsar is turned "on" and in the other it is turned "off." The other object in the pictures, a more ordinary star, continues to glow brightly.

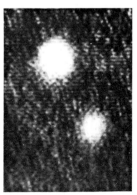

Above: A black hole is a region of intense gravity in space. It will draw down into itself anything that comes near, to be lost forever – such as the spaceship in the picture. But there is a strange contradiction in all this. As the spaceship is drawn towards the black hole, it speeds up faster and faster, but this means that time will pass more and more slowly. From the Earth (*our* time) it will seem as though the spaceship reaches the black hole and disappears. But to the spaceship crew (*their* time) it will look as though the black hole is getting no nearer – or even that they are never going to get there.

Pulsars

Some far-off starlike objects give out their radiation – light, radio waves or X-rays – in very short, very regular pulses. Some pulsars do this once every 4 seconds or so, but most pulse more rapidly – some many times each second. But just what is a pulsar and why does it pulse? The most likely answer has been provided by astronomers trying to work out what happens in a giant star explosion or supernova. Their calculations show that one of the things that can happen is the formation of a tiny but immensely heavy star called a neutron star.

Such a star, composed entirely of neutrons, or electrically neutral subatomic particles, would spin very fast. As it spun, it would give off pulses of radiation. It seems, then, that neutron stars and pulsars could well be the same thing.

Mysterious objects that appear in our skies, supposedly from outer space, are known as UFO's – short for Unidentified Flying Objects. Here is one over a house. It *could* be a visitor from far regions of space, of course; or it could be a special and rare effect of weather, dust and sunlight.

Quasars

The biggest of modern telescopes have revealed starlike objects at an enormous distance which are stranger still than pulsars. The farther away an object is in space, the faster it seems to be retreating, and quasars are moving away from us fastest of all. As they do so, they are giving off more energy than anything else in the Universe. Although they are so far away, the radiation we receive from them is greater than that from countless giant galaxies containing equally countless millions of stars. What, then *are* quasars? No one really knows the answer to this question, but many scientists believe that quasars represent young galaxies in the process of forming soon after the Big Bang.

Black Holes

After a supernova, the matter remaining behind may collapse inwards to become a super-dense neutron star. But if the exploded star was big enough, the matter will collapse even further under the enormous pull of its gravity – until matter has disappeared entirely, leaving only gravity itself. Such a strange region of space is called a black hole. As more and more matter is dragged into a black hole, more and more radiation will be given off. So quasars could be far off, super-giant black holes.

How Did it All Begin?

As astronomers began to discover the real size of the Universe, their ideas about it underwent several great changes. The first astronomers were stargazing priests. They explained the Universe in terms of gods and demons. Later on, the ancient Greeks had more logical explanations for what they observed in the night sky. The Universe, they said, is a harmony, or system, of perfect spheres, one inside the other. This idea lasted for nearly 2,000 years until Galileo, with the aid of his telescope, showed that it was hopelessly wrong. Bigger and better telescopes have showed modern astronomers more and more of the Universe, yet even with the most powerful modern telescope, no astronomer has yet seen all of the Universe. But, if we have yet to find out about the farthest limits of the Universe, can we hope to understand how it all began?

These two problems are in fact closely related to one another. As astronomers look deeper and deeper into space, they are also looking farther and farther back in time, towards the beginnings of the Universe. At present, the favored idea is that the Universe started with a gigantic explosion, 18,000 million years ago. This Big Bang Theory is illustrated by the pictures. A rival idea, out of fashion at the moment, says that the Universe has always looked more or less as it does now, but that new matter is being created all the time.

1

2

3

Right: As the Universe cooled down from the Big Bang, matter was first formed. When matter had cooled sufficiently, the formation of planets became possible. But even a planet, as this erupting volcano picture shows, can be very hot. Only when our planet had cooled down still further was life a possibility.

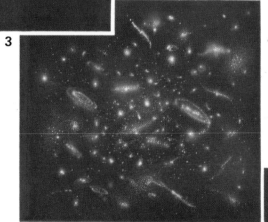

Above: The Big Bang Theory of how the Universe began. 1. The Big Bang itself. In the beginning, the entire Universe was compressed, at an unimaginably high temperature, into a ball. Eighteen thousand million years ago, this ball exploded. 2. The energy and matter of the Universe, cooling down after the explosion but still incredibly hot, spread widely in all directions. Atoms were first formed at this stage. 3. Millions of years later, the Universe began to look rather as astronomers see it through telescopes today. Matter had cooled down sufficiently to allow the formation of condensed forms such as galaxies, star clusters and cosmic gas clouds or nebulae. Perhaps even planets had been formed. 4. The Universe today. Galaxies and other cosmic bodies continue to fly apart from one another and the size of the Universe is now so great that no telescopes can penetrate to its farthest limits. Will the Universe continue to fly apart, forever? Or will it one day come to a halt, then begin to collapse inwards? Astronomers do not have the answers to these questions, but they hope to within the next century, when giant telescopes will have been sent into space.

4

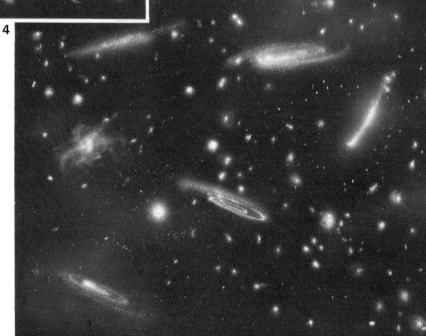

Using Telescopes

Even before the invention of the telescope, astronomers had done much useful scientific work, describing in great detail what they had observed in the night sky. They used scientific instruments, very like the sextants still used by sailors, to measure the apparent heights of the stars in the sky. By fixing the positions of the stars accurately the astronomers could make reliable star maps. The greatest of these early scientific astronomers was Tycho Brahe, who in the 16th century fixed exactly many hundreds of stars in this way.

Even the earliest and crudest telescopes opened a whole new Universe to the astronomer. The number of stars he had been able to observe with the naked eye was wretchedly limited because most stars are too faint to be seen directly. But even with the small magnification of the early telescopes, vast numbers more became visible. The early telescopes also revealed for the first time fascinating

Below: Even as long ago as the late 18th century, telescopes had become large and powerful scientific instruments. Biggest of all at this time was the reflecting telescope of Sir William Herschel, an English astronomer. Built in 1789, it had a magnifying mirror well over a yard wide. Using this and earlier telescopes, Herschel made many important discoveries, including the snows of Mars, the moons of Saturn, the outer planet Uranus, double stars and nebulae.

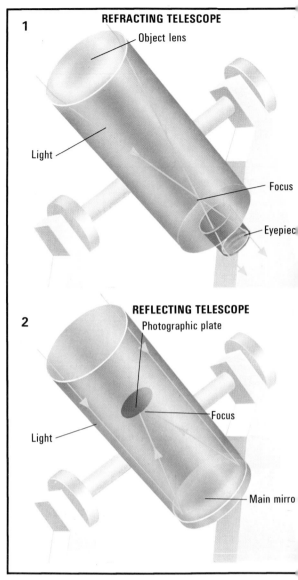

REFRACTING TELESCOPE
1
Object lens
Light
Focus
Eyepiece

REFLECTING TELESCOPE
2
Photographic plate
Light
Focus
Main mirror

Above: Optical telescopes are those that capture and magnify a light image of a visible object, so that the object looks nearer and more detailed. The diagrams show four kinds of optical telescope used in astronomy.

1. The refracting telescope is the type invented nearly 400 years ago and used by Galileo to make the first great discoveries of modern astronomy. It is called a refracting telescope because it captures and magnifies a light image with a large glass lens, called the *object lens*, which bends or refracts the light beams as they pass through it.

The astronomer views the magnified image through another, smaller, glass lens at the other end of the telescope, called the *eyepiece lens*. Refracting telescopes are still widely used, for example by amateur astronomers working from their own homes. Nowadays, the cheaper telescopes of this sort have lenses made of transparent plastics rather than the more expensive glass lenses.

2. The reflecting telescope was invented and first used by the great scientist Isaac Newton 300 years ago. Nowadays all the largest optical telescopes are reflecting telescopes. This type is called a *reflecting telescope* because it captures and magnifies a light image of an object by means of a large, curved mirror which reflects light, rather than by means of a lens which refracts light.

The mirror is more efficient than an object lens for magnifying light images. Newton with his reflecting

CASSEGRAIN TELESCOPE

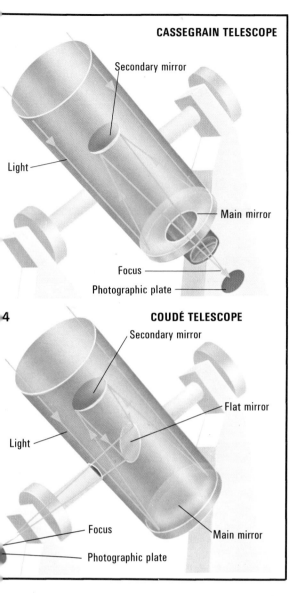

Secondary mirror

Light

Main mirror

Focus

Photographic plate

COUDÉ TELESCOPE

Secondary mirror

Flat mirror

Light

Focus

Main mirror

Photographic plate

4

Right: Early telescopes were often elegant instruments made from shining brass. This Cassegrain reflecting telescope was a portable instrument made in the 1790s by a famous London instrument maker, Jesse Ramsden.

telescope, like Galileo with his refracting telescope, viewed the magnified images through a small eyepiece lens. But in the telescope shown in the diagram, the eyepiece has been replaced by a photographic plate, which records a permanent image of the object.

3. The Cassegrain telescope was invented in 1672 by an astronomer of the same name. It, too, is a reflecting telescope which magnifies using a curved mirror. But in this case a second, smaller mirror is also used, which directs light back, through a hole in the center of the larger mirror, to an eyepiece or photographic plate. This method produces a larger and more detailed image than the Newton telescope.

4. The coudé telescope is yet another type of reflecting telescope. It is named after the elbowlike bend followed by the light beams passing through the telescope – *coudé* means "elbowed" in French. The coudé telescope uses two smaller mirrors in addition to the large magnifying mirror. As the telescope is moved to focus on different objects, the mirrors are turned automatically so that the light beams of the image are always directed towards the same point. This is necessary when the image is being recorded on a bulky photographic apparatus which cannot easily be moved around.

Modern giant optical telescopes, such as the ones shown on the next page, use both the Cassegrain and the coudé methods for recording astronomical images.

details of our own Solar System. They showed this to be only an insignificant fraction of the whole Universe.

The first telescope was invented in 1608 by a Dutch optician named Hans Lippershey. This was a *refracting* telescope (see diagrams) that used a glass magnifying lens to make objects look larger and nearer, just as most telescopes sold in stores still do today. In 1609 the great scientist and astronomer Galileo Galilei heard of this new instrument and set about devising his own, improved version. Over a period of several years he observed the night sky through his telescope and made many astonishing discoveries. He saw that the Moon has mountains. He studied the Sun and observed the sunspots on its surface (other astronomers also claimed to have discovered these). But as a result of gazing at the Sun, he suffered damage to his eyes and in old age he became blind.

As he observed the planets, Galileo discovered that Jupiter has four moons (12 more have been discovered since). He saw that Venus had phases, waxing and waning just like our own Moon. These and other discoveries led him to believe in Copernicus's idea of a Sun-centered Solar System. At the same time he began to think that the Universe might, after all, be infinitely large.

The refracting telescope was soon improved still more by another world-famous scientist and astronomer, Johannes Kepler. But this kind of telescope had one drastic limitation. Glass magnifying lenses could not be made very large, and so the power of such telescopes was strictly limited. The next really great improvement in telescopes was made by Isaac Newton who invented the *reflecting* telescope. This uses not a glass lens but a curved mirror to do its magnifying. Brightly silvered mirrors could be made much larger than glass lenses, so that Newton's reflecting telescope was a much more powerful instrument than Galileo's and Kepler's refracting telescopes.

Bigger and Better Telescopes

As we have seen, the telescope was invented at the beginning of the 17th century, when a glass lens was first used to do the magnifying. Then Isaac Newton invented the reflecting telescope, which uses a large mirror as its magnifier.

During the 18th and 19th centuries, bigger and better versions of both telescopes were built. Biggest and best of the 18th century was Herschel's reflecting telescope, shown on page 28. The biggest of all refracting telescopes was built at the Yerkes Observatory, Wisconsin, in the late 19th century. It is still in use today. It has a glass lens 40 inches wide, which is about as big as a lens can be. A larger lens would be so heavy that it would bend, or flow, and so distort the light that passed through it.

Giant Mirrors

Telescope mirrors, on the other hand, can be made considerably larger without this problem. In our own century, even greater telescopes have been built, all of which use great glass mirrors. One of the two biggest of all is shown by the pictures on this page. The Hale telescope at Mount Palomar, California, with its 16½-foot mirror, has been used to probe immense distances into space, to reveal galaxies and other luminous objects so far away that their light has taken thousands of millions of years to reach us.

All the big telescopes mentioned on these pages are situated on mountains. Good telescope viewing needs clear air, and air is clearest

Above: An astronomer looks through the eyepiece of the world's second biggest reflecting telescope, the Hale telescope on Mount Palomar, California. He is looking through the giant telescope at its coudé focus, which is shown below and explained in the diagram on page 29. If he wanted it, the part of the Universe he is examining could be recorded on a photographic film placed at the same focus.

Below: The Hale telescope. Its huge mirror is 16½ feet across and weighs 15 tons. (The biggest reflecting telescope of all, in the Caucasus mountains of Russia, has a mirror 19½ feet across.) The Hale telescope weighs in all more than 450 tons, yet it is so finely balanced that only a small motor is needed to turn it smoothly on its bearings. This motor is controlled by a computer, so that the telescope is turned to the precise direction required by the astronomer.

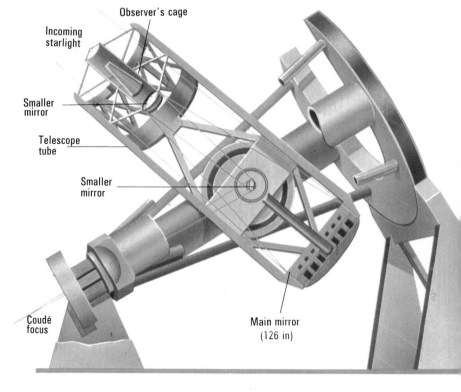

Observer's cage

Incoming starlight

Smaller mirror

Telescope tube

Smaller mirror

Coudé focus

Main mirror (126 in)

high up above the smog of modern civilization.

Telescopes in Space

Even better telescope viewing is obtained without any air at all, for the Earth's blanket of air is always on the move, distorting the light from any object in outer space before it reaches our eyes – this is the reason why the stars seem to twinkle.

Astronomers would rather do without star-twinkles, and so they are eager to put telescopes into space. This will allow them to make observations that are completely free from the problems of air distortion and of cloudy weather. With the coming of the Space Age, this can now be done, and the first space telescope is due to be carried into space by a shuttle during the 1980s.

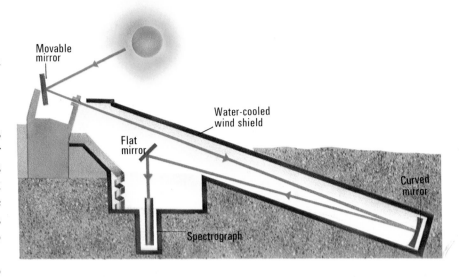

Above: A special kind of telescope has been invented for looking at the Sun in great detail. This solar telescope, like the other telescopes on this page, uses mirrors to reflect, magnify and focus the heavenly object being looked at (in this case, the Sun). A flat mirror at the top of the solar telescope catches the image of the Sun and reflects this down to the bottom of the tube. The flat mirror turns automatically, following the Sun precisely as it moves through the sky. At the bottom of the tube is a curved mirror which reflects and focuses the enlarged image of the Sun. A third, flat, mirror finally reflects the image on to a photographic plate. Because the Sun's rays carry heat as well as light, most of the very long telescope (over 49 feet altogether) is buried deep in the ground, which keeps it at an even temperature.

Giant reflecting telescopes are housed in great domes called observatories, like these at Mount Stromlo, Australia.

Radio Telescopes

We learn about stars and many other far-off heavenly bodies from the light they give out. This light may reach us only after long periods of travel across empty space. Nearly 400 years ago, the first telescopes made use of this light to provide more information about the visible Universe – just as modern optical telescopes do today. But what else besides visible light reaches us from outer space?

The answer to this question was not known until the late 19th century. At that time, radio waves were first discovered. These waves are the same sort of energy as visible light, but they have a longer wavelength and they are invisible to our eyes. However, as we all know, they are very useful indeed for carrying messages and entertainment around the world in the forms of sound radio and television. We produce these radio waves, but other radio waves were found to reach us from outer space.

Radio Stars and Galaxies

Radio waves were first traced to outer space about 50 years after radio waves were discovered. In 1931, a radio engineer in the United States named Karl Jansky was experimenting with some radio equipment when he discovered certain radio signals – rather crackly ones, like "static" – that he decided came not from Earth, but from outer space. By the 1940s, special telescopes had been built to receive these mysterious radio signals and to find out just where they came from. Radio telescopes, as they are called, are really giant aerials, or antennas, which serve the same general purpose as the antennas which receive television and radio signals for our everyday sets. The radio waves they pick up can be reproduced as a picture on a screen, just like a television picture. But radio telescopes are very much bigger than any television antenna, and are usually dish-shaped rather than H-shaped. Their great size and their special shape are both connected with the radio astronomer's need to collect lots of radio waves and at the same time to find out exactly which part of the sky they are coming from.

Using their giant new instruments, radio astronomers were able to draw up a new kind of picture of the Universe – a radio picture. Turning their radio telescopes on the Milky Way, they discovered that many radio sources, or broadcasters, existed there, in our own galaxy. What were they? One of these radio sources, for certain, was very close to us indeed – it was none other than the Sun. This showed that stars were one source of radio waves. The Sun is in fact quite a weak radio source, but because we are so close to it, we collect very large amounts of its radio waves. A much stronger source of radio waves in the galaxy was found, for example, in the Crab nebula, which is known to be the result of a great star explosion, or supernova (see page 24).

Searching still farther into space, well beyond the Milky Way, radio astronomers found far stronger sources of radio

Below: A radio telescope need not be a single, dish-shaped antenna: it can be a whole set of such antennas, grouped together to make a single great receiver of radio waves from outer space. This set of eight radio antennas makes up the radio telescope at the Mullard Radio Astronomy Observatory at Cambridge, England. The antennas are mounted on a railtrack, so that they can be moved closer together or farther apart, as required. At their farthest apart the antennas make up a single giant radio telescope 3 miles long.

waves. Their broadcasts reach us only after traveling through space for thousands or even millions of years. Some of these sources are other galaxies, which are such powerful broadcasters of radio waves that they are known as radio galaxies. Possibly even farther away, and giving out still greater amounts of radio waves, are the quasars. On the screen these look more like stars than like galaxies, yet they are so mysteriously remote from us that even radio astronomers are not at all sure what they are.

By building still larger radio telescopes, such as the giant Arecibo telescope shown below, radio astronomers hope to solve this and many other space problems. And, by sending special instruments into space, scientists can now detect not only radio waves and light coming from many parts of the Universe, but also invisible rays of a much *shorter* wavelength, such as X-rays and gamma rays. These methods, which enable astronomers to build up a "total wave picture" of the Universe, are shown on the next pages.

Below: The biggest single-dish radio telescope in the world is a two-tenths of a mile across. The huge dish of its reflector is placed in a natural bowl in the mountains at Arecibo, in Puerto Rico. Suspended 492 feet above the dish on three tall pylons is the radio antenna of the telescope. Unlike most other radio telescopes, the giant Arecibo dish is fixed in one position and cannot be steered to point various ways.

Windows
in Space

The very first astronomers were stargazers who lacked any scientific instruments for looking into space. Then telescopes were invented, which allowed astronomers to look much deeper into the Universe. These were optical telescopes, which capture and magnify optical, or light, images of stars and other celestial objects. So it can be said that a new, "optical window" had been opened on the Universe.

The optical window is still very important today, as shown by the giant modern optical telescope in the picture. But the picture also shows that the optical window on the Universe is a very narrow one. Light rays – the rays that we see by – are only a tiny fraction of the total amount of radiation that reaches Earth from outer space.

Other kinds of radiation that beam in continuously include rays that are both longer and shorter than those of visible light. Radio waves and infrared waves are longer, and gamma rays, X-rays and ultraviolet rays are shorter.. All these kinds of radiation are given out in abundance by one or another celestial object, and so provide very useful information about these objects. In our own century, science has perfected radio telescopes and many other kinds of instruments to detect and measure all these different kinds of radiation.

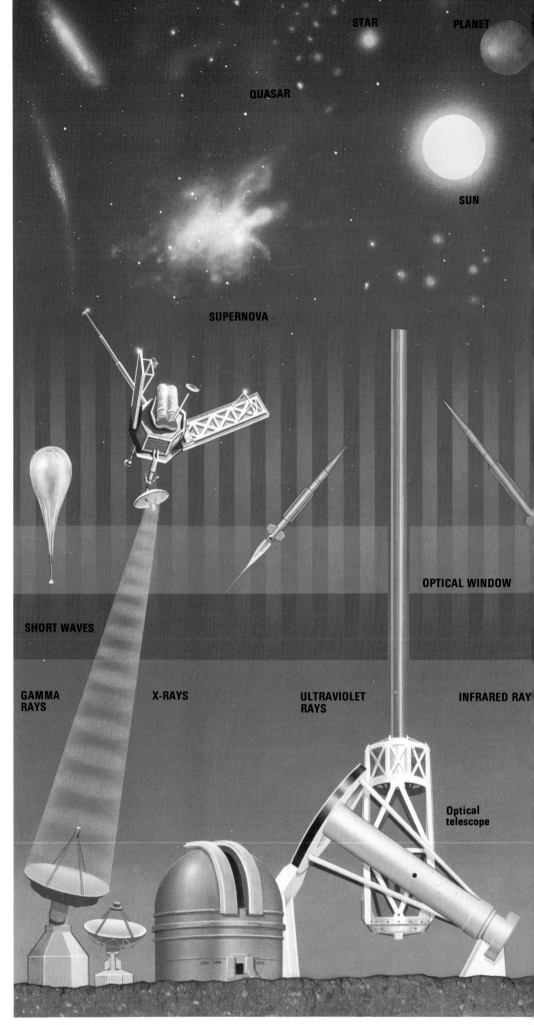

STAR

PLANET

QUASAR

SUN

SUPERNOVA

OPTICAL WINDOW

SHORT WAVES

GAMMA RAYS

X-RAYS

ULTRAVIOLET RAYS

INFRARED RAY

Optical telescope

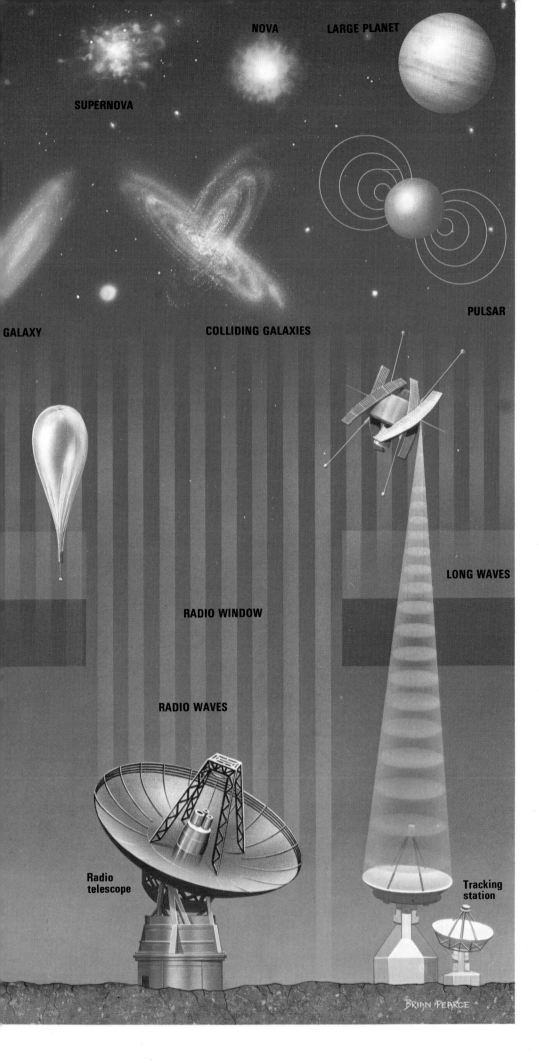

SUPERNOVA

NOVA

LARGE PLANET

GALAXY

COLLIDING GALAXIES

PULSAR

LONG WAVES

RADIO WINDOW

RADIO WAVES

Radio
telescope

Tracking
station

BRIAN PEARCE

The "radio window" in space is a much wider one than the optical window. Radio waves given out by stars and other celestial objects can be detected by radio telescopes on the ground, or are first trapped by satellites in space, then beamed down to a receiving antenna on the ground.

Many of the much shorter-length waves, such as gamma rays and X-rays, are stopped by the Earth's atmosphere. Scientists can get information about these short waves either by capturing them with a satellite above the atmosphere, or by sending up rockets or special balloons fitted with radiation detectors.

At the top of the picture are shown various kinds of celestial objects that give off one or more of these kinds of radiation. For example, as we all know, the Sun gives off light radiation and heat radiation (infrared rays). Much less well known are the mysterious, faraway objects called *quasars*, which give off huge amounts of radio waves.

35

A Guide to the Stars

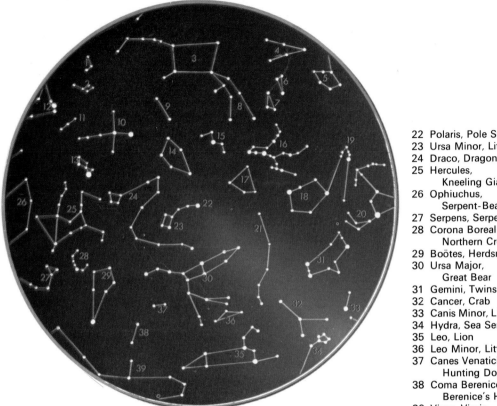

1 Equuleus, Colt
2 Delphinus, Dolphin
3 Pegasus, Flying Horse
4 Pisces, Fishes
5 Cetus, Sea Monster
6 Aries, Ram
7 Triangulum, Triangle
8 Andromeda,
   Chained Maiden
9 Lacerta, Lizard
10 Cygnus, Swan
11 Sagitta, Arrow
12 Aquila, Eagle
13 Lyra, Lyre
14 Cepheus, King
15 Cassiopeia
   Lady in Chair
16 Perseus, Champion
17 Camelopardus,
   Giraffe
18 Auriga, Charioteer
19 Taurus, Bull
20 Orion, Hunter
21 Lynx, Lynx

22 Polaris, Pole Star
23 Ursa Minor, Little Bear
24 Draco, Dragon
25 Hercules,
   Kneeling Giant
26 Ophiuchus,
   Serpent-Bearer
27 Serpens, Serpent
28 Corona Borealis,
   Northern Crown
29 Boötes, Herdsman
30 Ursa Major,
   Great Bear
31 Gemini, Twins
32 Cancer, Crab
33 Canis Minor, Little Dog
34 Hydra, Sea Serpent
35 Leo, Lion
36 Leo Minor, Little Lion
37 Canes Venatici,
   Hunting Dogs
38 Coma Berenices,
   Berenice's Hair
39 Virgo, Virgin

Star maps have been needed since ancient times, mainly for guiding travelers. Sailors particularly, traveling well out of sight of land, have long used the stars to guide their way. Since a clear sky shows such a bewildering variety of stars, ancient travelers found it helpful to give the brightest stars particular names so that these could be more easily told apart. Also, to make star maps it soon became necessary to arrange the stars into different groups so that any particular star could be picked out more quickly as a member of a particular group. These star groups are called *constellations*, and we still use the ancient constellations when describing the night sky.

Names for the Stars

The peoples of ancient civilizations, such as those of ancient Egypt and Mesopotamia, regarded the stars as visible evidence of gods, demons, heroes and mythical creatures that ruled the fates of mortal human beings here below. For this reason, the star groups or constellations were given the names and shapes of these supernatural beings.

Above: The constellations and stars of the northern hemisphere. Observers looking south will see the stars of the bottom left quarter of the circle in spring, the stars of the top left quarter in summer, the stars of the top right quarter in autumn and the stars of the bottom right quarter in winter.

As you can see from a glance at the lists of constellations on this page and those following, most of their proper names are not Egyptian nor Mesopotamian but Greek and Roman. This is because the Greek and Roman names have largely replaced those of the still older civilizations. The astronomer Ptolemy named 1,022 stars and 48 constellations. He lived in the second century AD in Egypt, in the city of Alexandria, which was famous as the greatest center of Greek learning. Since Ptolemy's time, another 40 constellations have been named, so that star maps of today contain up to 88 constellations. Many new constellations were named in the 17th and 18th centuries, at the beginnings of the modern scientific age. Then they were given the names of land animals, fishes, instruments and tools.

NORTHERN HEMISPHERE

Looking North

January

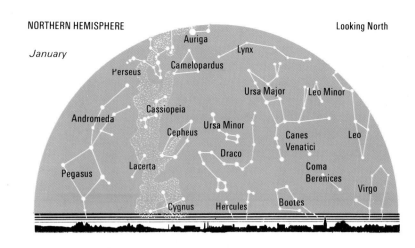

Right: Star-watchers of northern countries, looking north in the month of January, will see overhead a very bright star in the constellation Auriga, the Charioteer. This star is Capella, one of the three brightest stars in the northern hemisphere, and the sixth brightest of all stars.

NORTHERN HEMISPHERE

January

Looking South

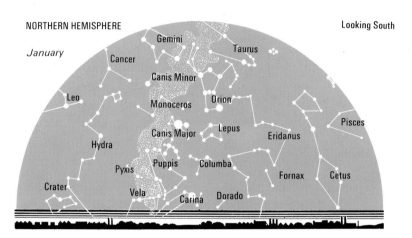

Left: Looking south in the January night sky, the same observers will see several glitteringly bright stars. One of these, Sirius, in the constellation Canis Major, the Great Dog, is the brightest of all stars in the heavens. (Note that Sirius and its constellation are also often visible in the southern hemisphere.) The next brightest star in the starscape is Rigel, seventh brightest of all stars, which appears in the constellation Orion, the Hunter. The second brightest star of Orion (in his "left shoulder") is the reddish star Betelgeuse, one of the greatest of all supergiant stars.

July

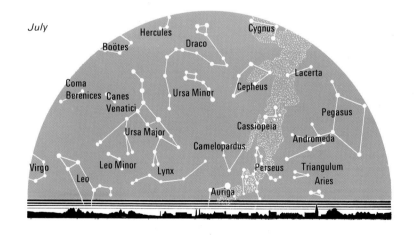

Right: In July, northern observers, looking north, will again glimpse the very bright star Capella, but now it has moved from its January overhead position to a point very low down in the sky. One famous constellation fully visible in both January and June is Ursa Major, the Great Bear, also popularly known as the Plow.

July

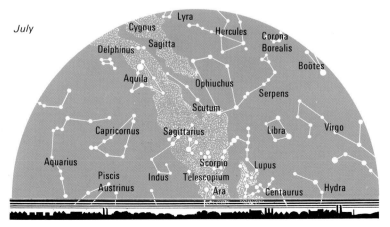

Left: Looking south in July, northern star-gazers will see one very bright star, Arcturus, that was not visible in January. Arcturus, the fourth brightest of all stars, is a red giant star. Its constellation, Boötes, The Herdsman, is only partly visible.

The Moving Constellations

When you look into the night sky, you do not always see the same stars and constellations. Even during a single night, the positions of stars and groups of stars will change. As the night goes on, you may observe one or two rise, or appear, above the horizon, while one or two more will set, or sink below the horizon.

Of course, you cannot hope to see all the constellations of the heavens, because the horizon – the line of the Earth's curvature – will always shut off a large part of the sky from your view. For this reason, people living in different places get a more or less different view of the night sky. As the star maps on these pages show, people living farthest away from one another – in the southern and northern hemispheres, respectively – see a very different set of constellations.

Wherever you live, the constellations that you do see appear to move during the night for the same reason that the Sun appears to move through the daytime sky. The movement is really one of the Earth turning on its own axis. But the constellations also change their positions (or appear to) in a second way. If you look at them carefully at different seasons of the year, say, at the same hour of the night, you will notice many changes. Any constellation that you saw before will have changed its position, while others will be missing from the sky. Yet other constellations will have newly appeared.

These changes, too, are caused by movement of the Earth, rather than by any actual shifting about of the stars. As the Earth moves in its yearly orbit around the Sun, anyone at a particular spot on the Earth's surface will get a changing view of the starry heavens. This explains why, at different times of the year, your own particular star map is so different.

The Zodiac

Ancient astronomers first noticed that the Sun, Moon and planets all moved in the same pathway through the sky. They also observed that at various times of the year, twelve constellations of stars also came into this pathway. These twelve are the constellations of the zodiac, four of which, The Crab, The Archer, The Bull and The Scorpion, are pictured on these pages. People who believe in astrology think that these twelve constellations have a magical influence on human lives.

Below: The constellations and stars of the southern hemisphere. Observers looking north will see the stars of the top left quarter of the circle in spring, the stars of the bottom left quarter in summer, the stars of the bottom right quarter in autumn and the stars of the top right quarter in winter.

CONSTELLATIONS OF THE SOUTHERN HEMISPHERE

1 Cetus, Sea Monster
2 Sculptor, Sculptor
3 Aquarius, Water-Bearer
4 Piscis Austrinus, Southern Fish
5 Capricornus, Sea Goat
6 Grus, Crane
7 Phoenix, Phoenix
8 Fornax, Furnace
9 Eridanus, River Eridanus
10 Hydrus, Little Snake
11 Tucana, Toucan
12 Indus, Indian
13 Sagittarius, Archer
14 Aquila, Eagle
15 Corona Australis, Southern Crown
16 Pavo, Peacock
17 Octans, Octant
18 Dorado, Swordfish
19 Pictor, Painter's Easel
20 Columba, Dove
21 Lepus, Hare
22 Orion, Hunter

23 Monoceros, Unicorn
24 Canis Major, Great Dog
25 Puppis, Poop
26 Carina, Keel
27 Volans, Flying Fish
28 Chamaeleon, Chameleon
29 Apus, Bird of Paradise
30 Triangulum Australe, Southern Triangle
31 Ara, Altar
32 Scorpio, Scorpion
33 Serpens, Serpent
34 Ophiuchus, Serpent-Bearer
35 Lupus, Wolf
36 Centaurus, Centaur
37 Crux, Southern Cross
38 Musca, Fly
39 Vela, Sails
40 Pyxis, Compass Box
41 Hydra, Sea Serpent
42 Sextans, Sextant
43 Crater, Cup
44 Corvus, Crow
45 Libra, Scales
46 Virgo, Virgin

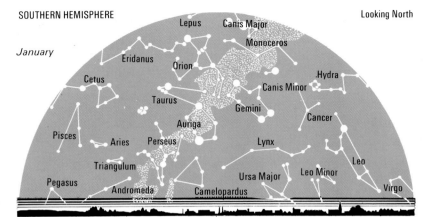

SOUTHERN HEMISPHERE

January

Looking North

Left: Observers of the January night sky in the southern hemisphere, looking north, will see a number of very bright stars that also appear in the skies of the northern hemisphere. Sirius, the dog star, brightest of all stars, shines directly overhead. Procyon in the constellation Canis Minor, the Little Dog, is the eighth brightest of all stars. Capella, sixth brightest, is visible in its constellation Auriga, much more in the middle of the sky than in northern latitudes.

Right: When the January observers of the southern latitudes look south, they see a number of exceedingly bright stars that never appear in skies of the northern hemisphere. Chief among these is Canopus, in the Constellation Carina, the Keel. Second brightest of all stars after Sirius, Canopus is a supergiant star, like Betelgeuse in Orion. Two more glitteringly bright stars appear in the southern constellation Centaurus, the Centaur. These are known as the Southern Pointers, because they point towards the constellation Crux, the Southern Cross, which itself contains a number of very bright stars. One Southern Pointer is the third brightest of all stars, while the other is the tenth brightest.

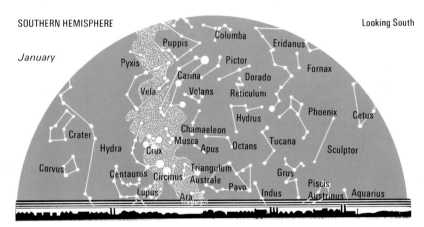

SOUTHERN HEMISPHERE

January

Looking South

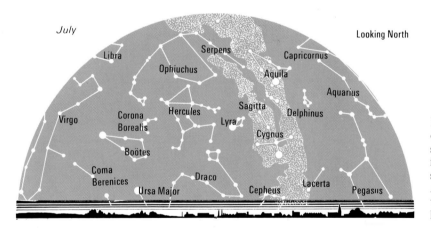

July

Looking North

Left: In the July night sky of the southern hemisphere, observers facing north will see the fourth brightest of all stars, the red giant Arcturus. This is also visible in July in the northern hemisphere, but in the southern hemisphere all of its constellation, Boötes, can be seen. Another major star visible in both hemispheres at various times is Vega, fifth brightest of all stars. This lies in the constellation Lyra, the Lyre.

Right: Dominant in the July night sky of the southern hemisphere, to observers looking south, are again the pair of Southern Pointers, pointing towards the four bright stars of the Southern Cross. To complete the named list of the ten brightest stars in the heavens, Achenar, ninth brightest of all stars, is visible at one end of its long constellation Eridanus (which was named after a river).

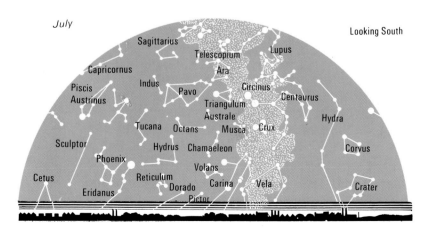

July

Looking South

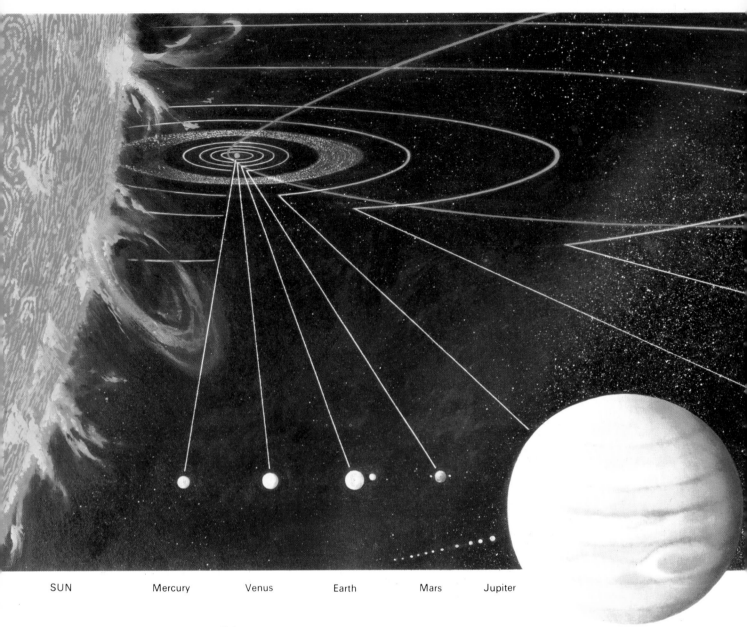

| SUN | Mercury | Venus | Earth | Mars | Jupiter |

The Solar System

The Sun and its family, the Solar System, consists of the Sun itself, an average-sized star, surrounded by a group of nine planets and their moons, together with a quantity of smaller asteroids and other space debris.

The Universe is so incredibly vast that it must contain untold numbers of other solar systems, but our own is the only one of which we have any real knowledge. Our knowledge has increased by leaps and bounds in the past 20 years, since the first age of space travel and exploration began. Space probes are now voyaging more than 621,371,000 miles from Earth, beyond the outer reaches of the Solar System, where our Sun is visible only as the largest of a host of stars against a dark sky.

How We See the Planets

The Sun provides our own planet Earth with life-giving light and warmth. It also lights up the other planets for us in the night sky. Some of the more distant planets are so faint that we need to use a telescope to see them. By contrast, Venus and Jupiter are so bright that they are widely recognized as the most brilliant starlike objects in the night sky.

Planets are easily told apart from true stars because they appear to move in a much more complicated, wandering pathway. This is because the planets are very close to us compared to the stars. What we see is really a combination of the planets' movement and that of the Earth, whereas the Earth's movement in space makes very little difference to the way we see the hugely distant stars move.

Saturn   Uranus   Neptune   Pluto

Origin of the Solar System

Scientists believe that the planets started as a spinning disk of cold gas and dust that formed around the young Sun, before it began to glow brightly. Gradually the dust and frozen gas began to collect together into lumps, which, in turn, joined up to form larger bodies. Eventually only nine large bodies remained, together with a number of smaller bodies, which became moons, and a certain amount of rubble.

As the Sun became hotter, the primitive atmospheres of the inner planets were boiled away and were replaced with gases released from inside. The outer planets remained frozen.

The Planets Today

The innermost and smallest of the planets is Mercury. It is so small and close to the Sun that it lost all its atmosphere. Its surface temperature is

hot enough to melt some metals. Next comes Venus, a planet that is continuously covered in dense clouds of acid vapor. It is even hotter than Mercury because its thick atmosphere of carbon dioxide traps heat like a greenhouse. The remaining inner planets are Earth, the only planet on which life has developed, and red, dusty Mars.

Outside these planets lies a belt of rocky lumps known as the asteroids. Then come the giants of the Solar System. Jupiter, famous for its red spot, is the largest of all. Saturn, with its thousands of rings, is also large. But it is much less dense than Jupiter and could, theoretically, float in water!

The outermost planets are Uranus, Neptune and the tiny Pluto. Uranus orbits the Sun at almost exactly twice the distance of Saturn. So when it was discovered in 1781 by William Herschel, the size of the known Solar System was immediately doubled.

Our Own Star

A clear night sky is full of stars, shining with a whitish, or sometimes a reddish or blueish brilliance. Because stars are so immensely distant from us, we cannot see their actual shapes, however large they may be. Only the light they give out in vast quantities, traveling perhaps for hundreds or even thousands of years through space, reaches our eyes at last as a faint or vivid twinkle.

Just one star is an exception to this very general description, and that is our own star, the Sun. We can see the shape of the Sun clearly because, relatively speaking, it is so close to us. Even the nearest star beyond the Sun is more than four light-years away – that is, its light takes more than four years to reach us. By contrast, the Sun is a mere eight light-minutes away – we see it as it actually was eight minutes ago.

An Ordinary Star

The Sun is exceptional among stars only in its nearness to us. In every other way, it is a very average sort of star, neither very large nor very small, neither very young nor very old, by star standards. Looked at from outside our own galaxy, the Sun would be lost in insignificance, being only one among the 100,000 million stars of its galaxy.

On the other hand, even a medium-sized, middle-aged star looks immensely impressive when seen close up, as we see the Sun every day. For one thing, the Sun is big enough to contain a million Earths. For another, the Sun is far hotter than anything we commonly experience here on Earth. Its surface temperature is 9,932°F – hot enough to boil any metal – and deeper down the Sun's temperature rises still higher to an unimaginable 27,000,032°F at the center.

The Sun seems to us to move through the heavens but of course it is we who move around the Sun. But the Sun, like all other objects in the Universe, does have its own movement. Like the Earth and other planets, the Sun turns on its axis, taking about 25 days to make one revolution. The Sun also journeys, together with its family of planets, in a great loop within its galaxy, the Milky Way. It makes this journey at a speed of nearly 497,097 miles per hour and makes one complete loop in 225 million years.

The Sun looks most impressive at sunrise or sunset, when for a short while we can look at it directly without being blinded by its glare, and when it looks larger than at other times. Both these effects are caused by the Sun's rays having to penetrate a greater distance through the Earth's atmosphere, when the Sun is low down in the sky.

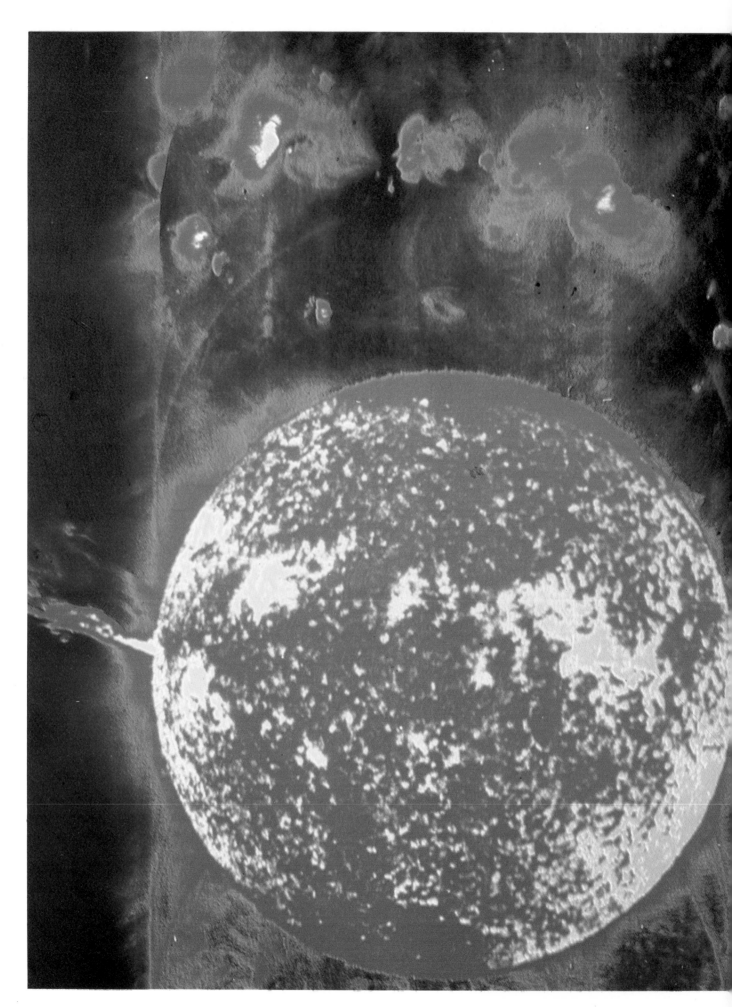

What is the Sun Made of?

Never try to look directly at the full Sun. Its light is dangerously strong. You can look briefly at the setting Sun, and then you will see a brilliant reddish disk. This surface area is called the *photosphere* (light-sphere). But the photosphere is not really the outermost part of the Sun. That can only be seen when the glaring disk of the Sun has been blacked out, as during an eclipse.

Then, around the edge of the photosphere, appears a brilliant, reddish and often very irregular layer. This is called the *chromosphere* (color-sphere). It is a zone of hot gas many thousands of miles deep, irregular because it is distorted by flamelike prominences and flares, which surge up from the interior of the Sun and then die down again.

Even now, the outermost layer of the Sun has not been reached. Outside the chromosphere is an even wider, but much paler, ring called the *corona*. Pearly white in color, the corona is also made of hot gas, but this is much less dense than the gas of the chromosphere, and eventually it thins out into space. Strangely enough, both the chromosphere and the corona are even hotter than the glaring yellow photosphere that is most visible to us. The corona, indeed, reaches a temperature of about 1,800,000°F.

The inside of the Sun also consists of gas, at ever-increasing temperatures. Below the visible photosphere, which is really a very thin and sparse layer, lie deeper and more concentrated layers of gas, which get hotter and hotter the farther down they are. The deepest, most central area is hottest of all and is the real powerhouse of the Sun.

Hydrogen and Helium

The Sun, then, is made of super-hot gas – and indeed at the star-temperatures of the Sun, any

Above: A peaceful, evening lakeside scene shows the Sun sinking towards the horizon, soon to be hidden by a line of trees. When the Sun is low, its glare is lessened, so that we are able to look at it without risking our eyes. At sunrise and sunset, we see the Sun through more of the Earth's atmosphere than at other times. This filters out many of the more dazzling rays, leaving mainly the less dazzling red rays. For this reason the Sun looks redder at these times. It also looks bigger, because the rays are refracted, or bent, outwards more by the atmosphere.

Left: This photograph of the Sun's disk was taken from Skylab. The surface layer of the Sun, called the *photosphere*, is at a temperature of 9,932°F – far lower than its center which reaches 27,000,000°F. From the surface erupt flamelike prominences, many thousands of miles high.

kind of solid or liquid matter would instantly boil off to become gas. The matter of the Sun is composed of the same chemical elements as those that make up the Earth and other planets. But there is one great difference. Nearly all the 92 chemical elements naturally present on Earth occur only in small amounts in the Sun. The Sun is made mostly of only two chemical elements. About 70 percent of the Sun's mass consists of hydrogen. But this is being steadily converted into helium (see page 71), which at present makes up 28 percent of the Sun's mass. So the Sun consists almost entirely of these two thin gases. All the other known elements make up only 2 percent of its mass.

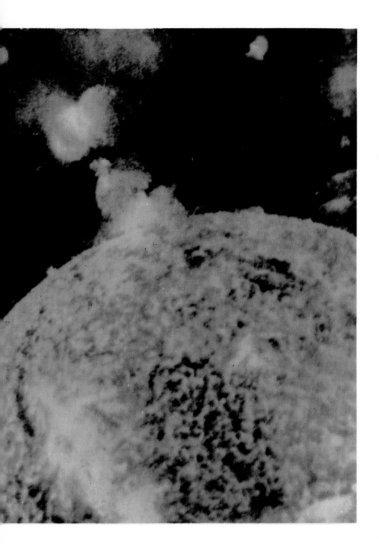

Left: Here on Earth the Sun seems to shine steadily, but when viewed closer up, our star is seen to have a very rough, or turbulent, atmosphere. This photograph was taken from Skylab in outer space and shows a huge solar prominence, or eruption from the Sun's surface. This giant eruption of super-hot gas reaches upwards from the Sun's surface for hundreds of thousands of miles – that is, a dozen or more times the width of the Earth.

We see the Sun, our parent star, as a blazing disk in the sky. It is dangerous to look directly at the Sun, because its fierce glare can easily damage our eyes. However, at dawn or sunset when the Sun is close to the horizon, there is more of the Earth's atmosphere between the Sun and us. It looks redder (and larger) and can be looked at for a while without harm. At all other times the Sun should only be inspected through a piece of dark glass which removes its harmful rays.

When we look at the Sun from here on Earth, unaided by a telescope, its disk looks peaceful enough, although a few small, uneven spots may be visible against it. Nothing could be farther from the truth than this apparent peacefulness. In reality, the Sun's surface and the climate above it are, compared with the Earth's climate, a raging inferno. The temperature at the Sun's surface is, of course, very much higher than any natural temperature on the Earth's surface: about 9,932°F. This is hot enough to turn anything here on Earth into a gas. So the Sun's surface, like its even hotter interior, is entirely gaseous. It is also very stormy indeed.

The Sun's Climate

Below: Solar prominences, or giant eruptions from the Sun's surface, take place over a matter of a few minutes to an hour or more. The prominence shown in the pictures lasted 12 minutes, from its first appearance, through its full mushrooming out from the Sun's surface, to its collapse and dying down. At its full height this prominence measured about 31,068.5 miles, or four times the width of the Earth. Although they are so spectacular, solar prominences are not easily visible from Earth, because the glare of the Sun's disk hides them. But they do become clearly visible, as in these pictures, when the Sun has undergone a complete eclipse, and its brilliant disk is hidden.

Solar Flares and Prominences

Seen from closer up, as in the top picture on this page, the Sun's surface is a swirling mass of hot gases. Below and between these surface swirlings, however, most of the Sun's surface can be seen to consist of an even, or regular, layer of extra-bright grains, shining out against a less bright background. Each of the extra-bright grains is about 620 miles across and lies close by its fellow grains, making a pattern of bright points. The grains are believed to be the topmost parts of columns of hotter gas coming up from the interior of the Sun. The less-bright areas between the grains are very likely cooler gas sinking down from the surface into the Sun.

But even these swirling hot gases are only the calmer aspects of the Sun's climate. Suddenly, one or more huge eruptions may occur, when fiery streams

of gas from the glowing surface shoot well into the corona or outer atmosphere of the Sun. These solar prominences, as they are called, are not only far larger than any storm effect here on Earth, but far greater than the Earth itself. The gas in a solar prominence is, like the rest of the Sun, mostly hydrogen. It is hotter than most of the remainder of the Sun's surface. What causes these giant prominences is not altogether clear.

Even hotter and faster outbursts from the Sun's surface are solar flares. These last only for a matter of minutes, whereas a solar prominence can last an hour. But solar flares, like solar prominences, contain vast amounts of energy. Radiant energy from a solar flare will often reach the Earth after about 24 hours. High up in the Earth's atmosphere, this solar radiation will cause the brilliant, multi-colored curtain of light called an aurora. And the energy from a solar flare will interfere badly with long-distance radio messages, because it causes electrical changes in the high layers of the atmosphere off which radio waves bounce.

Sunspots

Solar flares are connected closely with yet another type of giant disturbance of the Sun's atmosphere – sunspots. These are the darker spots that may be visible against the brighter disk of the Sun's surface. Unlike solar prominences and flares, sunspots are visible even when the Sun is not eclipsed (but *never* try to see them by looking directly at the Sun). They are darker than the rest of the Sun's surface because they are cooler, by as much as 3,600 degrees. They first appear as small, darker areas, grow for a few days into larger patches, then disappear slowly. Most sunspots last for a matter of days, although the largest of them, which can measure 124,274 miles across, can last for a month or more. What causes sunspots (and flares)? The answer is the tremendous magnetism of the Sun, which from time to time causes a local "dampening down" of the Sun's surface, the sunspot, together with a sudden increase in radiation, the flare. Sunspots, solar flares and some other irregularities of the Sun's surface reach a peak every 11 years. Again, we do not really know precisely why.

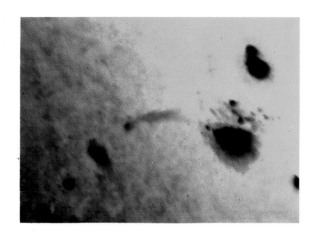

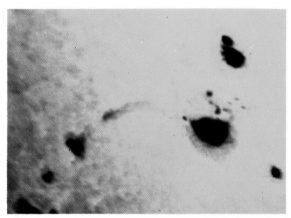

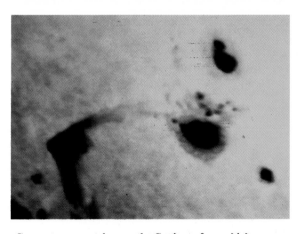

Above: Sunspots are patches on the Sun's surface which appear dark because they are at a lower temperature, only about 5,400°F to 7,200°F. They are caused by the Sun's magnetism, which causes at the same time the intense flashes of Sun-radiation called solar flares. In the pictures, a group of sunspots is shown going through a change of shape and area. This change, from top to bottom, lasted only about 30 minutes. One of the things that happened in this time was the ejection of a long streamer of hotter and more luminous gas.

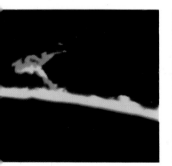
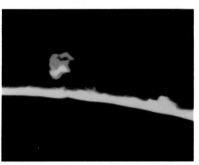
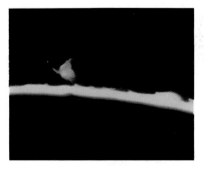
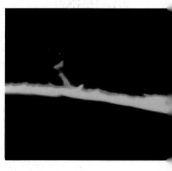

Planet Earth

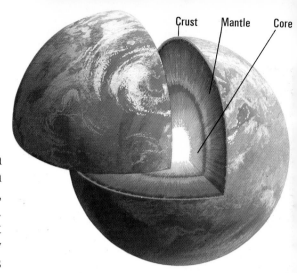

Our Earth is the third planet out from the Sun, and the fifth largest planet in the Solar System. It is much larger than Mercury, Mars and Pluto, much smaller than Jupiter, Saturn, Uranus and Neptune, and about the same size as Venus. Perhaps the Earth seems to be an average sort of planet. But space probes sent out to other planets in the last 20 years show that our planet is very probably unique in the Solar System, as it is the only planet that supports life.

One of the main reasons why Earth is so specially suitable for life is that it has an atmosphere rich in oxygen gas. All higher forms of life, as we know them, depend absolutely on this oxygen-rich atmosphere. Small planets, such as Mercury and Mars, cannot keep much of an atmosphere at all, because their forces of gravity are too weak to prevent atmospheric gases leaking away into space. Jupiter and the other giant planets have much higher gravities, and so are able to hold down very dense atmospheres. But their atmospheric gases are not the ones that can be relied upon to support any advanced forms of life. Perhaps surprisingly, the same is true of Venus, which, being about Earth-size, might be expected to have a similar atmosphere. Venus probes, however, show this planet to be a hot and hostile world swirling with dense, poisonous gases and with little free atmospheric oxygen.

In other ways, planet Earth is more like its fellow planets. Like them, it moves around the Sun in a path, or orbit, shaped like an ellipse. The time it takes for single orbit is 365 days (plus a few hours and minutes) which is the Earth-year. Its average distance from its parent star is about 93,205,650 miles.

Above: The Earth with a large slice cut out from it to show its several different layers. The outermost layer, or crust, is the one on which we live, and from which we get our mineral wealth. This crust is very thin compared with the other, deeper layers. The solid crust is a mere 5 to 19 miles thick, whereas the next layer down, known as the mantle, is no less than 1,740 miles thick. Like the crust, the mantle is made up of rocks, but these are under greater pressure and are very hot. Sometimes, rock from the upper mantle forces its way to the surface through cracks in the crust and pours as white-hot lava our of volcanoes. Deeper and wider still than the mantle is the Earth's core, which is about 4,300 miles thick. This core consists entirely or mostly of heavy metals. Very probably, it is a hot liquid on the outside but is solid, under enormous pressures, at the center.

Below: The tilt of the Earth's axis, as our planet orbits the Sun, explains the seasons of the year. The parts of the Earth's surface nearer the poles are tilted more towards or away from the Sun's rays and so have warmer and cooler seasons. The parts of the world nearer the Equator are tilted less from the Sun and so have a more even yearly temperature, with a more equal number of daylight hours.

THE SEASONS

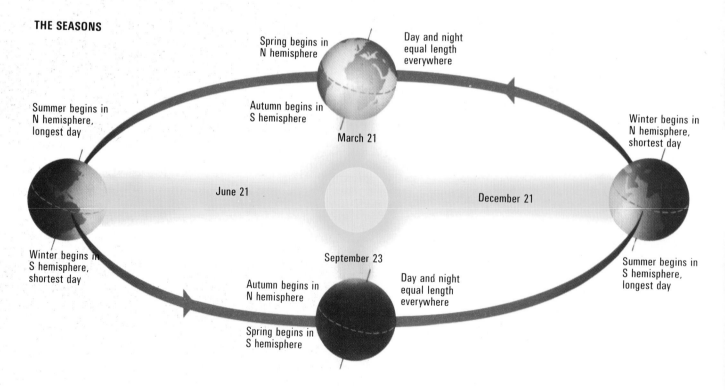

Spring begins in N hemisphere

Day and night equal length everywhere

Autumn begins in S hemisphere

March 21

Summer begins in N hemisphere, longest day

Winter begins in N hemisphere, shortest day

June 21

December 21

Winter begins in S hemisphere, shortest day

September 23

Summer begins in S hemisphere, longest day

Autumn begins in N hemisphere

Day and night equal length everywhere

Spring begins in S hemisphere

The Earth, like its fellow planets, spins on its axis, so that first one side, then the other, is turned towards the Sun. This gives us our night and day. A single spin takes just under 24 hours (to be exact, 23 hours, 56 minutes and 4 seconds) and so this is the length of the Earth-day, from dawn to dawn.

All the Sun's planets have axes which are tilted, when compared with the Sun's axis. The tilt of the Earth's axis is the cause of our warmer and cooler seasons, as shown by the diagram. And although Earth's surface is much cooler than those of its fellow inner planets, like them it has a hot, molten interior, as shown dramatically by the fiery eruptions of its volcanoes.

The Earth from outer space, photographed by astronauts of the Apollo 11 space mission, which later landed the first men on the Moon (see page 92). Part of the Earth is in shadow, just as part of the Moon is usually shadowed from us, here on Earth. More than half of the Earth's disk is, however, brilliantly lit by the Sun. Beneath the whitish clouds can be seen the dark area of the Mediterranean Sea, and the land masses and coastal lines of North Africa and Arabia.

Weather and Eclipses

The Sun is often said to provide "life-giving warmth," for without heat from the Sun's rays, all life would be impossible. Life cannot exist at very low temperatures. Life also needs the light in the Sun's rays. Without this sunlight, nothing green could grow on our planet. And without green plants, no animal life could exist because in the end all animals depend upon a constant supply of green food.

The very first green plants appeared on Earth about 2,500 million years ago. They were very small indeed, and were made up of only one or a few living cells. Such microscopic algae, as they are called, are still found on Earth today, living in vast numbers in rivers, lakes and seas. Like the larger plants with which we are more familiar, algae get their green color from a pigment called chlorophyll. This allows them to build up their own bodies using the energy of sunlight.

Sun and Atmosphere

On the primitive, lifeless Earth, the atmosphere was very different from the air we breathe today. It contained such unbreathable gases as hydrogen, methane and ammonia. It had little or none of the vital gas oxygen, which is needed by all higher forms of life. So how did the vital oxygen first get into the atmosphere? The answer is that it was

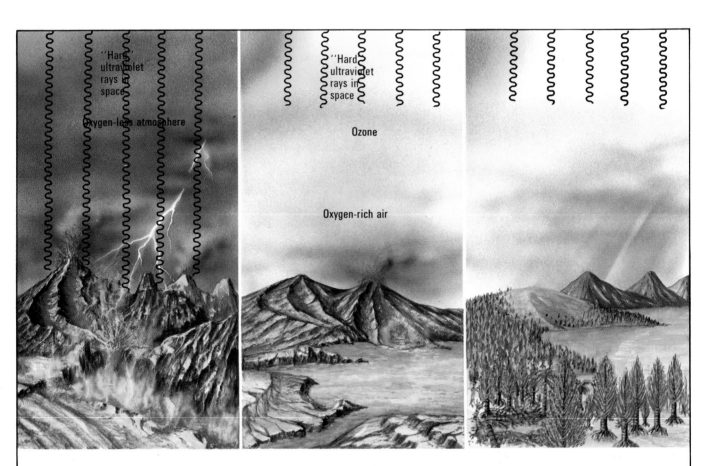

4,000 MILLION YEARS AGO
Weather has not yet formed on the primitive Earth. The hot, violent surface, bombarded with short-wave ultraviolet rays from the Sun, is perfectly lifeless.

1,000 MILLION YEARS AGO
Water has condensed on the cooler surface and primitive green life has arisen in it. This life gives off oxygen, which rises into the atmosphere. High up in the atmosphere the oxygen forms an ozone layer which blocks the lethal short-wave ultraviolet rays.

400 MILLION YEARS AGO
Life can now come to the surface, and eventually colonizes the land. Sun, wind, rain and frost slowly wear away even the highest mountains, so that much high ground becomes flatter. But other mountains arise due to movements of the Earth's crust – and are worn down in their turn.

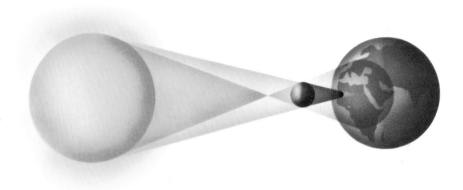

Above: When the Moon comes between the Earth and the Sun, an eclipse of the Sun occurs. The Sun's light is blocked out by the much nearer Moon, and the Moon's shadow falls upon that part of the Earth's surface that experiences the eclipse. Sometimes the eclipse is a *partial* one, when only a part of the Sun's disk is covered. The eclipse shown, however, is a *total* eclipse for the part of the Earth under the darker shadow. From there, the Sun will look as in the photograph, right.

The Sun is responsible for the weather on Earth, which shapes our planet. The Sun's heat lifts up the water that later pours down from rain clouds and collects in rivers, wearing away the rocks. The Sun's heat also provides the energy to make air circulate, and so is the real cause of winds which also cause erosion, or wearing away of the Earth's surface. This spectacular example of rock erosion is part of the Grand Canyon.

provided by the first tiny green plants, which gave off oxygen as they built up their bodies with the aid of sunlight.

At this early time, however, sunlight could kill as well as nourish, because it contained fatal amounts of ultraviolet rays. The first single-cell water dwellers escaped from these fatal rays by living deeper down, where the rays could not penetrate. The oxygen given off by these first plants rose up into the atmosphere. There some of it was changed by the lethal ultraviolet rays into another gas called ozone. The ozone, high up in the atmosphere, then blocked all further lethal ultraviolet rays from reaching the Earth's surface. In this way, life was able to come to the surface of the waters, and later to invade the land.

Sun and Weather
The Sun's rays are vital not only to life but also to weather. Heat from the Sun's rays evaporates moisture from the oceans, which forms the clouds that supply the land with life-giving rains. The Sun's warmth also causes the movements of air which we know as winds.

51

Our Neighbor, the Moon

The Moon is the nearest to us of all other worlds and, after the Earth itself, is the most familiar of all members of the Solar System. This is particularly true since astronauts first explored the surface of Earth's satellite and neighbor, a dozen or so years ago. Yet, centuries before anyone had landed on the Moon, astronomers had mapped its visible surface in the greatest detail.

These early astronomers gave the various regions that they identified Latin names, because this was the language in which scientists of all western nations then communicated with one another. They named many of the darker, more featureless parts of the Moon *maria* (singular *mare*) which is Latin for "seas." We now know that the Moon has no seas, and indeed no water at all. These large, smoother areas of the Moon's surface are really dusty plains lying between the equally bone-dry upland or mountainous parts of the Moon.

The Moon is a dead world, because no life can exist without water. The Moon is also entirely airless. The gravitational pull of the Moon is only one-sixth of Earth's, and this is not enough to hold air or any other gases near its surface.

As Apollo astronaut-moonwalkers discovered, the surface of the Moon can be exceedingly hot when lit by the Sun. The lunar night, however, is far colder than the coldest places on Earth. Yet scientists now believe that deep down inside the Moon there is a hot and still active core. This hot core would be the cause of the moonquakes – similar to earthquakes – detected and measured by scientific instruments set up on the Moon by the astronauts.

Possibly, the Moon was once a much more active world than it is now, with a hotter interior. Long ago, volcanoes might have poured out molten rock and fiery gases on to the surface, as they still do over parts of the Earth's surface. Such ancient volcanic activity can be held partly responsible for the Moon's most famous feature of all – its pockmarking of craters. But more certainly still, very many of these craters were formed in a quite different way, by the bombardment of meteorites from outer space.

MOON DATA

Diameter or width is 2,160 miles, or less than $\frac{1}{3}$ of the Earth's.
Mass or weight is only about $\frac{1}{80}$ that of the Earth.
Gravity is only $\frac{1}{6}$ that of the Earth, which is why Apollo Moon explorers could make those large leaps in their heavy space suits.
Surface temperature in sunlight is about 210°F, hot enough to boil water. Surface temperature of the lunar night is minus 238°F, twice as cold as anywhere on Earth.
Average distance from the Earth during its orbit, is 238,855 miles.
Orbiting time around the Earth is $27\frac{1}{3}$ days. But this is also the time the Moon takes to rotate once on its own axis, so that we always see the same side of the Moon from Earth.

The Moon, and Earth, have been bombarded with debris ever since they were formed, about 4,600 million years ago. On Earth erosion has removed all trace of meteorite impacts, but no erosion occurs on the airless, waterless Moon. Its jagged mountains are the result of a massive bombardment that went on as the Moon formed. When this had died down, lava welled up from under the crust, spreading out to form the "seas" or *maria*. Since then meteorites have continued to bombard the Moon, giving it its pockmarking of craters. Some of the craters, however, were formed by volcanic action.

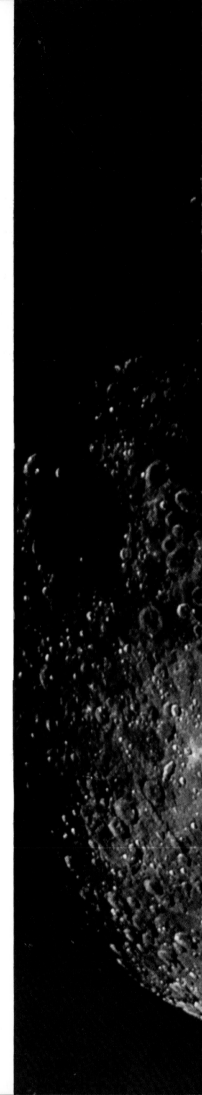

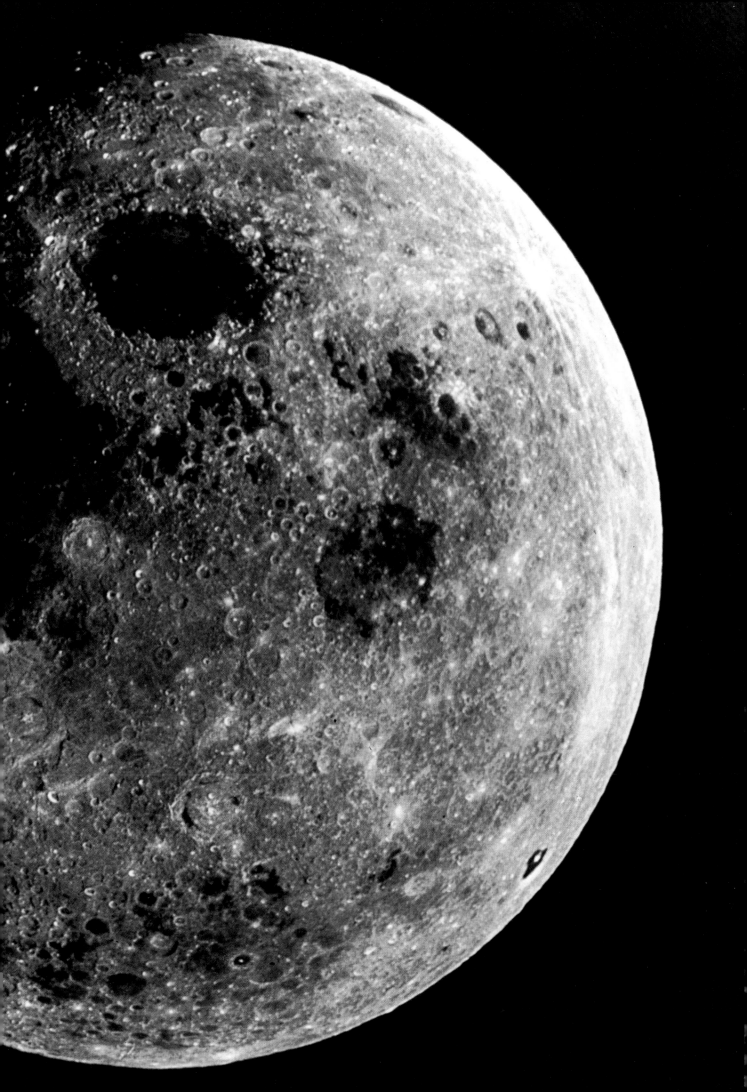

Sun, Moon and Earth

The Moon is a small world compared with its neighbor Earth, which weighs 81 times as much. But the Moon is big compared with nearly all the moons of the other planets. Sometimes, because they are nearer in size, Moon and Earth are known as a double planet.

Phases of the Moon

Everyone is familiar with the appearance of the Moon in the night sky. The different phases, or bright areas, of the Moon are the areas lit up by the Sun as the Moon travels around the Earth and these areas are reflected to our eyes. All parts of the Moon not reflected to our eyes are quite invisible in the night sky, because the Moon has no light of its own. When the lit-up area is increasing, we say that the Moon is waxing, and when it is decreasing, we say that the Moon is waning. This is shown by the diagram. Sometimes the Moon is made visible to

us not by sunlight reflected directly from the Moon, but by sunlight reflected from the Earth itself on to the Moon. This happens in the evening at the time of the spring new moon, when a "ghost" of the full moon is seen in the sky, lit up by earthshine.

Whatever the lit-up area of the Moon, it is always the same side of the Moon that we are looking at. This is because the Moon takes the same time – $27\frac{1}{3}$ days – to turn once on its axis, and to revolve once about the Earth. The first time that anyone on Earth saw the "far" side of the Moon was in 1959, when the Russian space probe Luna 3 traveled around the Moon and sent back pictures to Earth.

The Moon in the night sky looks different to us at different times of the month because it reflects more or less light from the Sun. These changes in lighted-up area of the Moon are called its phases.

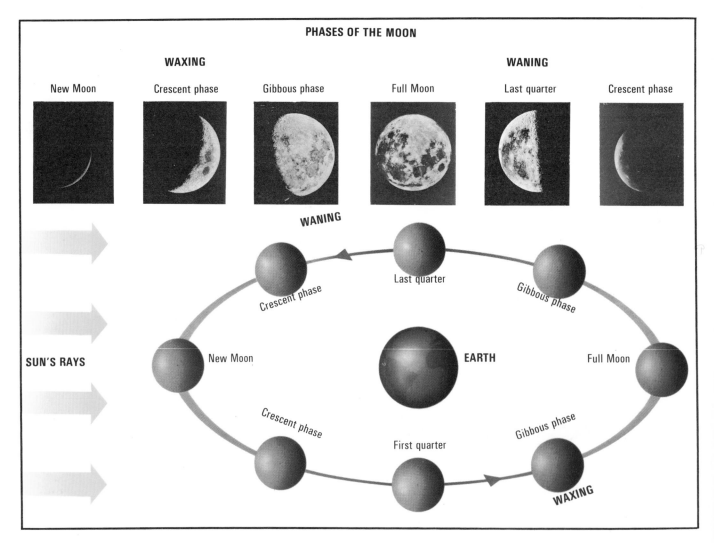

PHASES OF THE MOON

WAXING   WANING

New Moon   Crescent phase   Gibbous phase   Full Moon   Last quarter   Crescent phase

WANING

Last quarter

Crescent phase   Gibbous phase

SUN'S RAYS   New Moon   EARTH   Full Moon

Crescent phase   First quarter   Gibbous phase

WAXING

The time from new moon to new moon is about $29\frac{1}{2}$ days, more than two days longer than the time it takes to circle the Earth. This is explained by the fact that as the Moon revolves around the Earth, the Earth is traveling around the Sun. So the Moon takes about two days more to come into the same position, or angle, with regard to Earth and Sun.

Tides

As Isaac Newton discovered about 300 years ago, objects attract one another with a force of gravity; and the more massive and nearer the objects, the greater is the force drawing them together. As we have noticed already, the Moon is (for a satellite) a large world. And in astronomical terms it is quite close to the Earth. So we can expect the Moon to exert quite a strong pull on the Earth, and the other way around.

The Moon's pull on the Earth is most noticeable by the way it helps to cause the ocean tides. Water on the side of the Earth nearest the Moon is attracted and drawn up by the Moon's force of gravity, to make the bulge that we call a high tide. At the same time, the Moon's force of gravity can be considered to pull the Earth slightly *away* from its water on the far side – so that there is a bulge of water on both near and far sides of the Earth (see diagram).

However, the picture is a little more complicated still, because not only the Moon but also the Sun will attract the Earth's waters and so help to make its tides. The Sun, of course, is far more massive than the Moon, and so has a far greater gravity. But it is also far more distant from the Earth, so that it exerts a smaller pull on the Earth's waters. The highest tides occur when Sun and Moon are lined up on one side of the Earth, pulling together. The lowest high tides occur when the Sun and Moon are at right angles to each other.

But what about the movement of the tides across the oceans, which the pull of gravity, by itself, does not explain? We must not forget that the Earth is all the while spinning on its axis. As it does so, the solid Earth spins rather more quickly than its watery covering, which is held back by the pull of Moon and Sun. In any one place, we see a rise and fall of water as the Earth turns on its axis, which we know as the movement of the tide.

Both the Moon and the Sun help to produce the bulges in the Earth's ocean waters that we call the tides. The Moon, being very much nearer, does most of the pulling to make the bulge. The size of the bulge depends on the particular positions of the Sun and Moon with regard to Earth.

1 When Sun, Moon and Earth make a big angle with one another, then the Sun and Moon are pulling in different directions on the Earth's waters. For this reason, the bulge produced by the Moon will be less. The lowest high tides of the year are called *neap tides*.

2 When Sun and Moon are lined up to one side of the Earth, then they pull together on the Earth's waters, to make the biggest bulge. The highest high tides of the year, called *spring tides*, then occur.

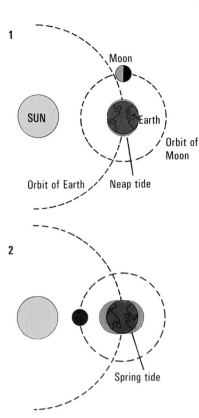

When the Earth comes between the Sun and the Moon, its shadow is thrown onto the Moon, so darkening it or causing an eclipse. The diagram on the right shows how the Earth's shadow can fall on the Moon. When the Earth is directly between Sun and Moon, the deepest part, or "umbra," of its shadow falls on the Moon. The photographs below show how the Moon looks during this *full* eclipse of the Moon. But when the Earth is less directly between the Sun and the Moon, only the less deep part, or "penumbra," of its shadow falls on the Moon. The Moon is darkened less than during a full eclipse, and is said to undergo a *partial* eclipse.

Mercury and Venus

Mercury, the closest planet to the Sun, is a hot little world about one and a half times the width of our Moon. Like that of the Moon, its surface is heavily pockmarked with craters, although it lacks the largest of the dark areas or *maria* of the Moon.

Mercury orbits its parent star at an average distance of 36,039,518 miles or only about one third the distance of Earth from the Sun. This closeness makes Mercury difficult to observe in the night sky, although sometimes it can be seen low down on the horizon. When Mercury passes *across* the Sun, it can been seen more clearly as a dark dot – although of course, no one should try to look for Mercury in this way, without using proper protection from the Sun's glare.

On its sunlit side, Mercury is so hot that metals such as lead and tin remain molten. On its dark side, the little planet is fatally cold. Such hostile conditions are likely to discourage astronauts from ever wanting to set foot there. The planet has the added disadvantage of being quite airless, because any gases escaping from its surface are quickly lost to space under its feeble force of gravity. But although Mercury is no world to live on, it is still very interesting to scientists, and space probes have been sent to the little planet to take close-up photographs and make other detailed measurements (see page 104).

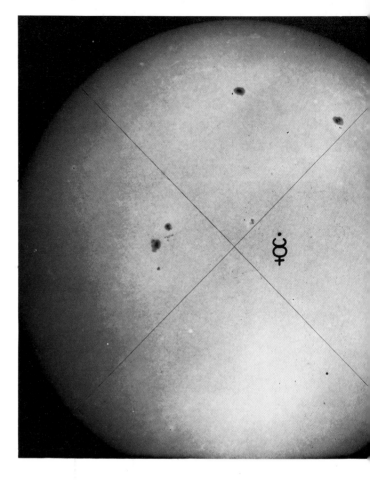

Above: Mercury and Venus are the two members of the Sun's family nearer the Sun than our own planet. For this reason it is sometimes possible for us to see one or the other of these planets against the Sun, as it passes across the Sun's bright disk. In this photograph, taken through dark glass to cut down the Sun's harmful glare, the little planet Mercury is seen as a dark dot. The other small dark areas are sunspots.

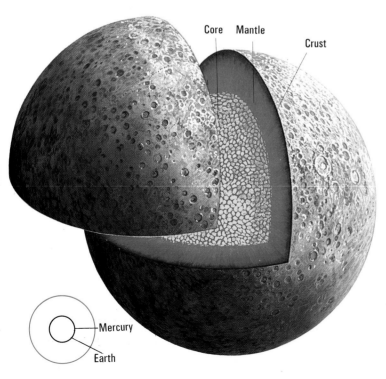

MERCURY FACTS	
Average distance from Sun	36,039,518 miles
Nearest distance from Earth	27,961,695 miles
Temperature on sunlit side	About 700°F
Temperature on dark side	About −328°F
Maximum diameter (width)	3,032 miles
Volume	$\frac{1}{16}$ that of the Earth
Mass, or weight	$\frac{1}{16}$ that of the Earth
Gravity	$\frac{1}{3}$ that of the Earth
Atmosphere	None
Number of moons	None
Mercury year (time of orbit around the Sun)	About 88 Earth days
Time taken to turn on axis	About 59 Earth days
Direction of rotation (turning on axis)	Counterclockwise – the same as Earth's

Left: A picture of Mercury with a slice taken out to show the planet's inner appearance. Although one of the smallest of the Sun's family, Mercury is also one of the heaviest for its size. Scientists believe that this heaviness is caused by a large core of hot iron. The airless surface of Mercury is riddled with craters, like that of the Moon.

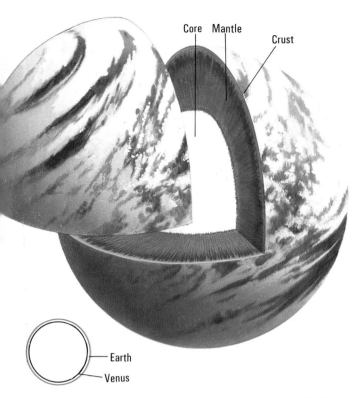

Core Mantle Crust

Earth

Venus

Above: This cutaway picture of Venus shows the probable inner details of Earth's sister planet. Although its surface conditions are very different from those of Earth, Venus is of about the same size and weight as our own world, and for this reason scientists guess that it may have roughly the same kind of interior.

Venus

The second planet out from the Sun, and nearest to ourselves, is Venus. It has often been called Earth's sister planet. It is only slightly smaller than Earth, and has about the same gravity. Venus is a brilliantly familiar object in our night skies, being known by the names of the Morning Star and the Evening Star (according to the time of its appearance); it is the brightest of all heavenly bodies other than the Sun and the Moon.

Through a telescope, Venus proves to be covered with cloud so dense that none of its other surface features can be seen. For a long time this thick atmosphere, apparently so like Earth's and so unlike the arid, airless surfaces of Mercury and the Moon, was thought to be a sign that Venus could hold life. But then it was discovered that the atmosphere of Venus is in fact suffocating and poisonous, and so totally unsuitable for the forms of life typical of Earth. Venus space probes (see page 104) have sent back information which discourages still further any idea of Venus as the fertile, swampy world that people once imagined.

Venus orbits the Sun in about 225 days but turns on its axis only once in every 243 days. The strange result of this is that a Venus day, from sunrise to sunset, lasts several Earth years.

VENUS FACTS

Average distance from Sun	67,108,068 miles
Nearest distance from Earth	26,097,582 miles
Surface temperature	About 860°F
Surface pressure of atmosphere	90 times that of Earth
Atmosphere?	Mostly carbon dioxide gas, with clouds of acid vapor
Maximum diameter (width)	7,543 miles (only 383 miles less than Earth)
Volume	$\frac{9}{10}$ that of Earth
Mass, or weight	$\frac{4}{5}$ that of Earth
Gravity	$\frac{9}{10}$ that of Earth
Number of moons	None
Venus year (time of orbit around the Sun)	Nearly 225 Earth days
Time taken to turn on axis	243 Earth days
Direction of rotation (turning on axis)	Clockwise – opposite to Earth's

Right: This vivid photograph of Venus was taken by the American space probe Mariner 10, in February, 1974. It shows clearly how heavy cloud swirls across the face of the planet, hiding all other surface details from view. The clouds also hide the Sun from the surface of Venus – and unlike Earth's clouds, these are made of a poisonous, acid vapor.

Mars

Mars is the next planet beyond Earth in the Solar System. Its red color led the ancient Greeks to identify it with their god of war. Much later, people began to imagine Mars as the home of alien forms of life. In *The War of the Worlds*, the most famous story about Martians, H. G. Wells made his alien creatures invade our own world and attempt to destroy its civilization.

As telescopes improved, Mars became a favorite with astronomers, who were fond of drawing maps of its surface. In the mid-19th century, the astronomer Giovanni Schiaparelli drew up the most detailed Mars map of all, showing a great network of canals stretching across the red planet's surface. This certainly looked like the work of engineers capable of giant achievements. In our own century, it has gradually become apparent that Schiaparelli's Martian canals are almost as much the work of imagination as H. G. Wells' intelligent-octopus Martians.

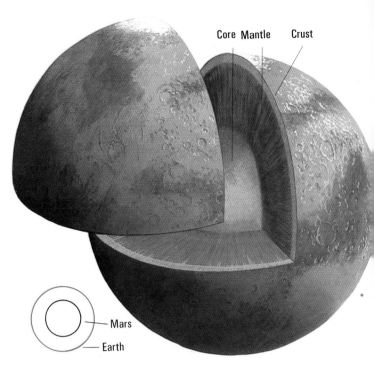

Above: A cutaway picture of Mars shows a large, hot core, perhaps of molten iron. Such a core could help to explain the volcanoes on the surface of Mars, some of which may still be active. Mars's surface is also pockmarked with craters, many of which will have been caused by the smashing impact of meteorites – Mars has little atmosphere to help burn these up before they reach its surface.

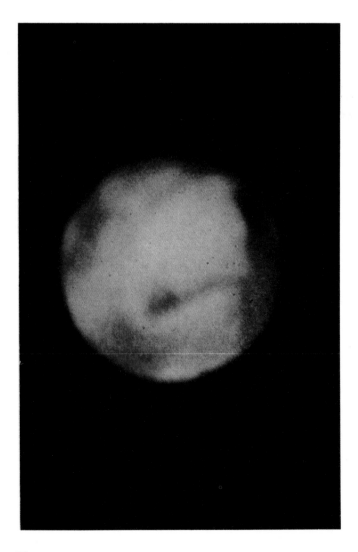

MARS FACTS	
Average distance from Sun	141,700,000 miles
Surface temperature	About freezing (32°F) in sunlight, down to minus 212°F or even lower in the polar night
Atmosphere	Very thin carbon dioxide gas
Maximum diameter (width)	4,200 miles
Volume	About $\frac{1}{7}$ that of Earth
Mass, or weight	About $\frac{1}{9}$ that of the Earth
Gravity	Nearly $\frac{1}{2}$ that of the Earth
Number of moons	2
Mars year (time of orbit around the Sun)	687 Earth days
Mars day (time taken to turn on axis)	24 hours $37\frac{1}{2}$ minutes – just over one Earth day

Facts about Mars

At the same time, astronomers were gathering a lot of really factual information about Mars. About one and a half times as far from the Sun as Earth, Mars is a colder world than our own. It is also much smaller, being about half the width of Earth and twice the width of the Moon.

The 19th-century astronomers could be fairly sure about one sort of marking on Mars's surface – the Martian polar caps. These almost certainly

Left: Through a powerful telescope, Mars appears clearly as a reddish world, but its markings are deceptive, as many mistaken theories about them show.

showed colder areas, perhaps of snow and ice like the poles of our own planet. In recent years Mars space probes have confirmed this idea, but they have shown that the polar water-ice of Mars is mixed with still colder dry ice, or frozen carbon dioxide.

The rest of the surface of Mars is nearly, if not quite, barren of water – even as ice. Mars is, compared with Earth, airless; it has only a very thin atmosphere of carbon dioxide, held to the planet's surface by a weak force of gravity. White clouds are often visible over the Martian surface, but these appear to be clouds of dust rather than of water vapor like our own clouds. Great dust storms are known to sweep at tremendously high speeds across the cold, airless deserts of Mars, and these, combined with its other disadvantages, make Mars a much more hostile world to life than had once been hoped by astronomers and others.

If Schiaparelli's canals have any foundation in fact, this would be in their faint resemblance to the long, deep, rift valleys that crisscross the eastern half of the globe of Mars. On Earth, rift valleys are caused by cracks opening in the Earth's crust because of movements farther down. But such valleys can also become wider as the result of rivers that wear them away. This hints that Mars may once have possessed water in the forms of rivers, lakes and seas – but even if so, all these must have dried up hundreds or thousands of millions of years ago. The largest rift valley is the enormous canyon known as Coprates. This is 2,485 miles long, 75 miles wide and 19,685 feet deep. Another spectacular feature on the surface of Mars is the gigantic volcano known as Olympus Mons. It was active until about 100 million years ago and now stands 15 miles high and measures 373 miles across its base.

Well above the surface of Mars orbit its two tiny moons, Phobos and Deimos. In 1971, a Mariner space probe (see page 104) showed them to be nothing more than irregular lumps of rock. Phobos is the larger of the two, measuring 17 miles across its widest part. It orbits Mars at an average distance of about 3,730 miles. Deimos measures only 9 miles across and orbits Mars at a distance of just under 12,427 miles.

These topographic maps of the surface of Mars were made with data compiled by the U.S. Geological Survey. Each color corresponds to an elevation interval of 1,640 feet. The views are centered on the equator at intervals of 90° of longitude and show what enormous progress astronomers have made in their knowledge of the planet.

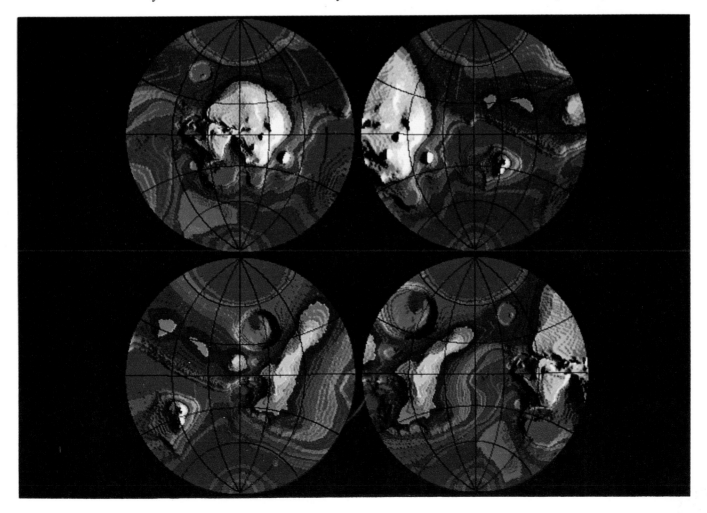

Jupiter

Jupiter is the giant of the Sun's family, a planet large enough to contain 1,300 Earths. The outer planets, Saturn, Uranus and Neptune, are also huge planets compared with our own, but Jupiter is so large that it would swallow these and the rest of the Solar System – with the exception of the Sun itself – about one and a half times over.

Jupiter is so big that if it had been much bigger it would have become not a planet, but a star. Like the Sun and other stars, Jupiter consists mostly of hydrogen. On our own planet hydrogen is a gas, but on Jupiter it is mostly a liquid. This is because all matter on the giant planet is compressed, or dragged inwards, by a tremendous pull of gravity. Such great compression will turn even hydrogen into a liquid. At the very center of Jupiter, the pressure is so great that the hydrogen may have turned into a metal-like solid, or it may still be liquid, but at a temperature far higher than that of any liquid on Earth – about 54,000°F. This temperature, although it is more than five times that of the Sun's surface, is still far from hot enough for the center of a star – which requires temperatures of millions, not thousands, of degrees. So Jupiter remains a planet, but one which makes so much heat from its own huge gravity that it sends out three times as much heat as it receives from the Sun.

Jupiter lies in the middle of the Solar System, being the fifth planet out from the Sun and also the fifth inwards from Pluto, the outermost planet. Jupiter is, on average, about five times farther out than Earth. For this reason the Jupiter year (the time taken for it to orbit the Sun) is also much longer than the Earth year; in fact, it is about 12 of our years.

Jupiter is never much closer to the Earth than it is to the Sun. The nearest that the giant planet gets to us is about 404 million miles. As it is so very large, Jupiter reflects a great deal of light from the Sun, which makes it brilliantly visible in our heavens as one of the brightest of "stars."

Jupiter's easy visibility has made it a great favorite of astronomers. The first astronomer to make scientific discoveries with a telescope, Galileo, quickly turned his instrument onto Jupiter and proved that it was a planet, not a star – which upset all the ideas then held about the Universe (see page 18).

Much more recently, astronomers of our own times have begun to uncover many more of Jupiter's secrets with the aid of giant telescopes and other modern scientific instruments. Most recently of all, space probes have been sent out from Earth, to pass close by the giant planet and its many moons. You can read about these latest discoveries on page 108.

JUPITER FACTS

Average distance from Sun	483,700,000 miles
Maximum diameter (width)	88,607 miles
Volume	1,316 times that of Earth
Mass, or weight	About 318 times that of Earth
Gravity	More than $2\frac{1}{2}$ times that of Earth
Type of atmosphere	Hydrogen and other gases
Number of moons	16 (so far discovered)
Jupiter day (time taken to turn on axis)	9 hours 55 minutes
Jupiter year (time of orbit around the Sun)	Nearly 12 Earth years

Jupiter has been called a gas giant because it consists mostly of hydrogen, which at Earth temperature is a gas. However, because of Jupiter's tremendous pull of gravity, nearly all its hydrogen and other matter is present in liquid form, although the giant planet does have a gassy atmosphere. At the center of Jupiter, the enormous pressure creates a temperature several times hotter than that of the Sun's surface, and Jupiter gives out more heat than it gets from the Sun. Space probes have recently shown that Jupiter is surrounded by a faint ring.

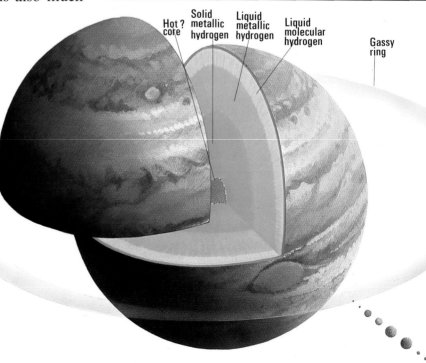

Hot? core — Solid metallic hydrogen — Liquid metallic hydrogen — Liquid molecular hydrogen — Gassy ring

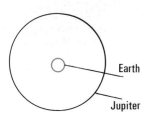

Earth

Jupiter

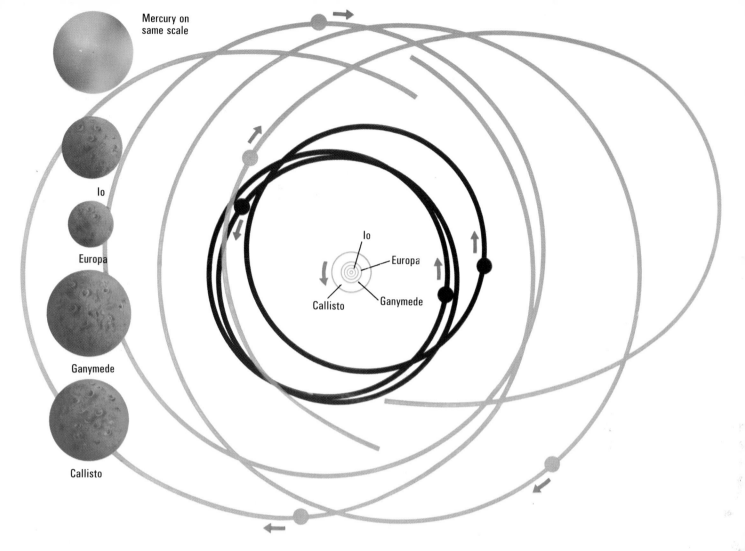

Mercury on
same scale

Io

Europa

Ganymede

Callisto

Io
Europa
Callisto
Ganymede

Above: This diagram shows Jupiter's four largest
moons compared with the smallest planet of the
Sun's family, Mercury. Jupiter's 16 or so moons
are in three groups, separated from one another
by great gaps in space. The innermost group is the
one containing the four large satellites. The next
moons are much farther away from their parent
planet, but orbit it in the same direction. The
outermost group is much farther away still, but its
moons travel around Jupiter in the opposite
direction. To complete this complicated moon
picture, two of the smaller moons have orbits
which change in direction every time around –
shown by the broken orbit lines. Not every one of
Jupiter's moons is shown, because this would
make the picture more complicated still.

The two giant planets Jupiter and Saturn both
have a numerous family of satellites. Jupiter has
about 16 moons, of which four are about the size
of our own Moon, or of a small planet. Here five
pictures taken by the Voyager spacecraft have
been combined to show Jupiter and its four largest
satellites.

61

Saturn

Saturn, the planet famous for its rings, is the second largest planet of the Sun's family. Across its equator, or widest part, it measures 74,564 miles. However, the width of its rings increases this to more than 169,000 miles. In the night sky, Saturn appears as a bright star, but like other planets it can easily be told from a true star by its more complicated motion across the heavens. We cannot see Saturn's rings with the naked eye, but they will be visible with a telescope of moderate size unless the rings are end-on to us, as in the picture for 1951 at right. The rings are so thin compared with the rest of the planet, that it takes a powerful telescope to show them when they are in this end-on position.

Saturn is, on average, about nine and a half times farther out from the Sun than our own planet. It is also nearly twice as far out as Jupiter, and so receives much less heat from the Sun than that world. This may explain certain differences between the two giant planets. For example, special light-analyzing instruments show astronomers that Saturn's surface contains more liquid ammonia and methane than Jupiter's surface. Very likely, this is because more of these gases have condensed into liquids on Saturn's colder surface. But Saturn, like Jupiter, is made up mostly of hydrogen. It is much less dense than its larger fellow planet, being only as heavy as 95 Earths, whereas Jupiter is as heavy as 318 Earths. As with Jupiter, the hydrogen at the giant planet's center must be under such great pressure that it is a super-dense liquid or even a metallic solid.

Right: In its 30-year orbit around the Sun, Saturn displays its surface area and rings in many different aspects. The pictures show how Saturn appeared from Earth over a period of 14 years, from 1951 to 1975. In 1951, when the rings were end-on to us, a powerful telescope was necessary to make them visible at all, because they are so very thin compared with the planet itself.

Below: One of the Voyager photographs of Saturn.

1951

1960

1967

1975

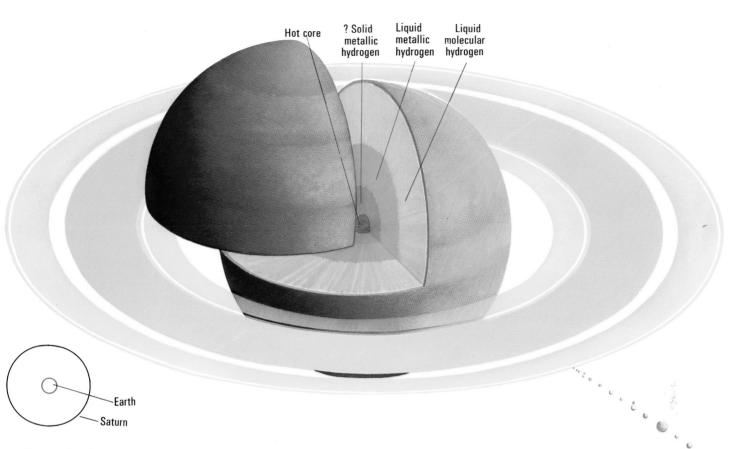

Hot core   ? Solid metallic hydrogen   Liquid metallic hydrogen   Liquid molecular hydrogen

Earth

Saturn

Saturn's rings

The rings of Saturn have been known ever since Galileo first turned his early telescope on the planet. But Galileo had only an imperfect view of Saturn and was frankly puzzled by its rings, which seemed to change from one viewing to the next. We now know that this puzzling change came about because the giant planet tilts more or less at different times, so that its rings become more or less visible.

The rings remained a great mystery even for later astronomers, who were very undecided about the number of rings there were and about what they were made of. The question of the precise number of rings is only now being answered, as space probes fly past Saturn and send back close-up, detailed pictures of all the rings and other fascinating features of the giant planet (see page 111).

The exact make-up of the rings also continued to be a mystery until very recently, although for about a century astronomers have thought that the rings must be made up of many small particles close together. This ideas was prompted by calculations which showed that if the rings were entirely solid, they would soon be pulled inwards by Saturn's force of gravity. The Voyager space probes have indeed shown that Saturn's rings are made up of vast numbers of small, glittering particles, perhaps mostly of ice or ice-covered rock – although just where these particles came from is still guesswork.

Above: A cutaway view of Saturn, showing the inside of the giant, lightweight planet. Three main rings can be seen through telescopes on Earth, but Voyager spacecraft show that Saturn's rings are made of hundreds of strands only a few miles wide, some one within the other and others braided together.

SATURN FACTS	
Average distance from Sun	886,696,417 miles
Maximum diameter (width)	74,564 miles
Volume	755 times that of Earth
Mass, or weight	95 times that of Earth
Gravity	About $1\frac{1}{4}$ times that of Earth
Type of atmosphere	Hydrogen ammonia and methane gases
Number of moons	24 (so far discovered)
Saturn day (time taken to turn on axis)	10 hours 14 minutes
Saturn year (time of orbit around the Sun)	$29\frac{1}{2}$ Earth years

The Outer Planets

Beyond the giant ringed planet Saturn stretches a great gulf of empty space, 932 million miles deep, until the next planet is reached. This is Uranus, which, on average, is nearly 2 billion miles distant from the Sun. If you have keen eyesight you may just be able to see Uranus in the night sky, looking like a very faint star. Through a telescope, it appears as a greenish disk, rather featureless except for a few dark streaks. Uranus is a large planet, its greatest width being nearly 32,200 miles, or about four times that of the Earth and less than half that of Saturn. Like Saturn it is a "gas giant," consisting mostly of hydrogen, and its greenish color is probably that of liquid methane and ammonia condensed on its cold surface. Uranus, like Jupiter, has rings of ice or rocky particles; these are smaller than the rings of Saturn and are very faint indeed.

One of the most remarkable features of Uranus is its tilt towards the Sun, which is greater than that of any other planet. Uranus takes 84 Earth years to orbit the Sun, and its strange tilt means that each of its polar regions has a night lasting 21 years. Fifteen moons have been discovered orbiting Uranus, but unlike the moons of Jupiter and Saturn, all these of Uranus are smaller than Earth's Moon.

Neptune

Neptune, second farthest out of all the known planets, is on average nearly 3 billion miles distant from the Sun. Although so much farther out, Neptune is practically a twin of Uranus. It is about the same size, being 30,758 miles wide, and is also a gas giant with a bluish or greenish color.

Being such a large planet, Neptune quite naturally has some effect, by means of its pull of gravity, on its near-twin Uranus. It was in fact the pull of Neptune on Uranus that led to Neptune's discovery. Neptune, being so distant, is never visible to the naked eye, and so astronomers understandably thought that Uranus was the last planet of the Sun's family. Then two astronomers noticed that Uranus had rather a peculiar orbit around the Sun, as though something was pulling it out of its true path. They guessed that this something was another planet, and at once set about calculating its position in the Solar System. A later astronomer confirmed this excellent guess, and the calculations, by discovering Neptune (as the new planet was then named) through the telescope. Later still, two moons of Neptune were discovered, one smaller than the Earth's Moon, and the other larger.

Little can be seen of Neptune's surface, even with the most powerful telescopes, but it is most probably an even colder version of the surface of Uranus. Voyager 2 space probe is now on its way to Neptune. It will reach the icy planet in 1989 and should tell us much more about this far-away member of the Sun's family.

Pluto

Still farther out beyond Neptune is Pluto, the most distant of all the Sun's planets. But the orbit of

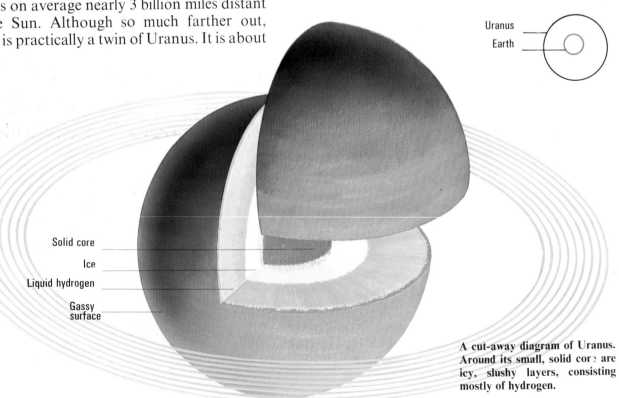

Uranus

Earth

Solid core

Ice

Liquid hydrogen

Gassy surface

A cut-away diagram of Uranus. Around its small, solid core are icy, slushy layers, consisting mostly of hydrogen.

Pluto is the most eccentric of all the planetary orbits, and sometimes Pluto may actually be *nearer* the Sun than Neptune at its farthest-out point. At the other extreme of its orbit, Pluto reaches the immense distance of more than 4 billion miles from the Sun.

Pluto, unlike the other outer planets, is no giant. It is about the size of the Earth's Moon, with a width of about 1,860 miles. Its small size, together with its peculiar orbit, gives a clue to the possible true nature of Pluto. This rocky little planet may once have been one of Neptune's moons, that escaped the pull of gravity of its parent planet to go off on an eccentric path of its own.

Unwarmed by its far-away parent star, Pluto is very probably so intensely cold that no gas on its surface can remain unliquefied. It turns on its axis in about six and a half of our own days, and takes more than 248 Earth years to complete its path around the Sun. Pluto is, of course, quite invisible in our skies, and even the most powerful telescope will show it only as a faint "star." Pluto was truly discovered by accident. After the discovery of Neptune, astronomers began to wonder if there were any more planets in the Solar System. The orbits of Uranus and Neptune still seemed to suggest the presence of another large planet. Over a period of about 80 years astronomers tried to calculate where such a planet would be. Pluto was finally discovered in 1930 when Clyde Tombaugh noticed an unknown object moving across the sky. But it turned out not to be the expected gas giant. In fact Pluto is so small that it cannot have any noticeable effect on Uranus and Neptune.

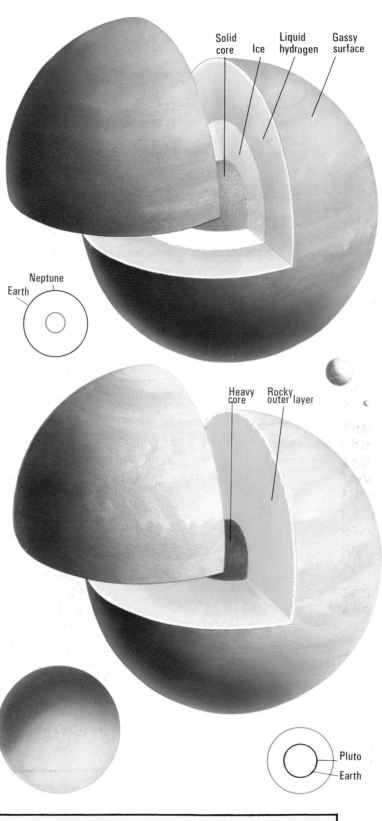

Above right: Neptune is much the same size as Uranus and its interior, too, is probably very similar.

Right: Pluto, the farthest-out of all the Sun's known planets, is a much smaller world than the gas giants, being only about the same size as our Moon. It is so faint, even through a giant telescope, that none of its details are visible. Possibly it is a rocky planet, with a heavier core.

OUTER PLANET FACTS			
	Uranus	**Neptune**	**Pluto**
Average distance from Sun	1,739,838,800 miles	2,796,169,500 miles	3,666,088,900 miles
Maximum diameter (width)	32,187 miles	30,758 miles	1,450
Volume	52 times that of Earth	44 times that of Earth	About $\frac{1}{20}$ that of Earth
Mass, or weight	About $14\frac{1}{2}$ times that of Earth	About $17\frac{1}{4}$ times that of Earth	Only a small fraction of that of Earth
Gravity, at surface	A little less than that of Earth	A little more than that of Earth	Only a small fraction of that of Earth
Type of atmosphere	Hydrogen, helium	Very cold gases	Probably none
Number of moons	15 (so far discovered)	2 (so far discovered)	One
Day (time taken to turn on axis)	About 16 hours	About 19 hours	About $6\frac{1}{2}$ Earth days
Year (time of orbit around Sun)	84 Earth years	About 165 Earth years	About 248 Earth years

Wanderers in Space

The orbits of three comets are shown in the picture. Encke's comet has a remarkably small orbit, coming very close to the Sun once every 3.3 years. Halley's comet voyages farther out, but not beyond the Solar System. We see it when it approaches the Sun closely, once every 76 years. Kohoutek's comet has been reported only once, in 1973. It will be about 75,000 years before it will be visible from Earth again.

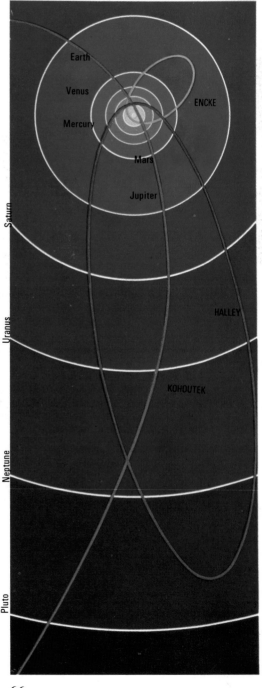

The frozen, dark world of Pluto orbits our Solar System at its farthest extent. Nearly 4 billion miles away, on average, from its parent star, this is for most of the time the remotest solar outpost. The empty space that stretches away from it to the next star beyond the Sun is far greater still. But, in fact, this immense tract of space *does* have inhabitants which belong occasionally to our Solar System. These are the comets, wanderers in space that may travel 6 trillion miles or more from the Sun, before curving back in their long, narrow orbits.

When a comet enters the Solar System, and gets relatively near the Sun, it can look quite spectacular to us on Earth. A comet such as Halley's Comet, which returns regularly every 76 years, can be relied upon to become one of the most brilliant objects in the night sky, for a short while at least. It is easy to see its brightly glowing, fireball-like head, and long, luminously streaming tail. A comet's tail may stretch for many hundreds of millions of miles.

Yet, outside the Solar System, where most comets spend the vast majority of their time, they are invisible to us. In empty space, they lack entirely that glowing head and tail and so are totally insignificant objects. Why this great change in their near-to and far-away aspects? Comets are, in fact, rather a cosmic confidence trick. Despite their brilliant appearance and the vast amounts of space they take up in the sky, in terms of solid matter they amount to very little: a comet is nothing much more than a rather small amount of frozen gas and water, mixed with dusty solid particles. One astronomer has described the substance of a comet as "a dirty snowball." It is only when this snowball gets near to the Sun that it begins to turn into a firework.

This drawing shows a comet seen in November and December 1680. It comes from a pamphlet entitled "An Alarm to Europe" describing "some preceding and some succeeding causes of its sad effects to the East and North Eastern parts of the world."

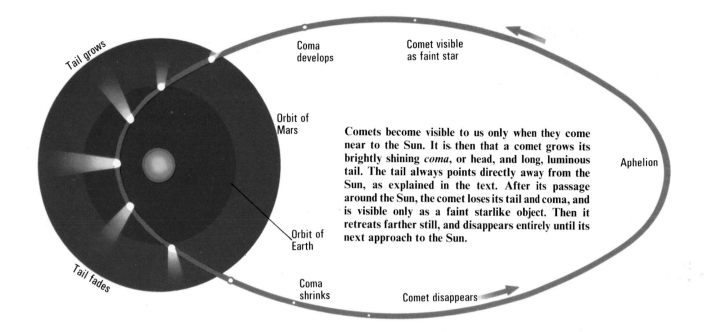

Tail grows

Coma
develops

Comet visible
as faint star

Orbit of
Mars

Aphelion

Comets become visible to us only when they come near to the Sun. It is then that a comet grows its brightly shining *coma*, or head, and long, luminous tail. The tail always points directly away from the Sun, as explained in the text. After its passage around the Sun, the comet loses its tail and coma, and is visible only as a faint starlike object. Then it retreats farther still, and disappears entirely until its next approach to the Sun.

Orbit of
Earth

Tail fades

Coma
shrinks

Comet disappears

How a Comet Gets its Tail

When a comet comes nearer to the Sun, its frozen packet of ice, dust and gas begins to melt. This releases some of the gases, which form a sort of halo around the head of the comet. This halo reflects sunlight and it can become very bright. At this point we first notice the comet as a moving dot in the night sky. But to form the comet's long, luminous tail, something very different then happens.

The ever-glowing mass of the Sun is releasing all the time and in all directions an invisible stream of particles. These are far smaller than the dusty particles of the comet. In fact, they are smaller even than atoms, being *subatomic* particles called protons and electrons. The great, radiating stream of particles from the Sun is known as the solar wind.

When the solar wind meets the head of a comet, it drives from it some of its gassy substance, which then streams backwards from the head, to form the comet's gassy tail. The electrons and protons of the solar wind are electrically charged particles, and their electrical charges cause the gases of the comet's tail to glow brightly. The tail always points directly away from the Sun, regardless of the direction in which the comet is moving. It is now that we see the whole comet – a sight which terrified people in the Middle Ages who believed that God was sending them a dire message.

What Happens to Comets

Vast numbers of comets circle the Sun, but at such great distances that we never see them. Only the occasional one, with a narrower orbit, comes close enough to be lit up for us by the Sun. Once in a while, however, a comet approaches *too* close to the Sun. Then it breaks up under the Sun's gravity, and disappears for ever.

Humason's comet was discovered in 1961. It will not be seen again for 2,900 years. The stars appear as streaks because they moved during the time the camera took the photograph.

Below: Ikeya-Seki comet is one of the brightest to be discovered this century, but in this picture it lacks a head. It was seen in 1965, and will reappear in 880 years.

Debris in Space

The Sun's family of planets was formed at about the same time as the Sun itself. This happened about 5,000 million years ago, when the Sun and planets condensed, or formed, out of a great cloud of cosmic gas. Most of the gas cloud went to form the Sun itself. Other large amounts condensed to form the giant planets such as Jupiter. Lesser amounts formed the medium-sized planets such as Earth, and still lesser amounts made the small planets such as Mercury.

But the Solar System also contains members that are far smaller even than the smallest planet. These are the asteroids, or planetoids, that sweep in a large band or belt between the orbits of Mars and Jupiter. The largest asteroid, Ceres, has a diameter of only 621 miles compared with 3,032 miles for the smallest planet, Mercury. Most of the asteroids – and there are 100,000 of them – are far smaller still, being mere lumps of rock coated with ice.

Astronomers first thought that this debris of the Solar System was the result of a planet breaking up early in its life. Yet there seemed no particular reason why a planet should have broken up, and the modern idea is that the asteroids were formed at the same time as the rest of the Solar System. They are in fact rubble, which never at any time went towards making a planet.

Above: The vast majority of meteorites never reach the Earth's surface. They are so small that they burn up entirely in the atmosphere. A few larger ones reach the Earth, however, including this iron meteorite, which would have been visible as a fireball for thousands of miles if it landed at night. This meteorite has been sawed in half, then polished and etched with a strong acid, to reveal its characteristic crisscross pattern.

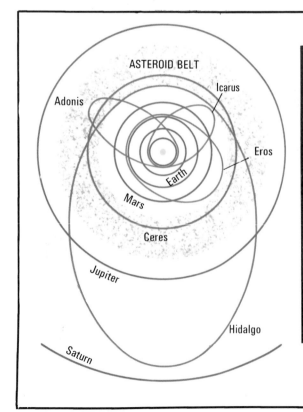

The diagram on the left shows the orbits of five of the bigger asteroids, namely Ceres, Icarus, Eros, Adonis and Hidalgo. You can see that these orbits, like those of the host of smaller asteroids, lie mainly between the orbits of the planets Mars and Jupiter.

Asteroids are tiny compared with stars or even with the planets and are particularly difficult to see in the night sky. But because they are very close to us, compared with the stars, they can sometimes be spotted by their greater movements. In this photograph an asteroid appears among the stars as a short streak of light.

Meteors and Meteorites

Asteroids are much smaller than planets, but most meteorites are smaller again than asteroids. Meteorites are lumps of metal and rock that sometimes enter the atmosphere of the Earth. Indeed, there are so many, many meteorites that on a clear night, they can be seen falling through the sky at the rate of about five an hour. Then they are known as meteors, or shooting stars.

The quick streak of brilliance that is a shooting star represents the burning up of a tiny meteorite as it plunges through the Earth's atmosphere. Long before it can reach the Earth's surface, such a trifling bit of space debris has heated up and burned away in a puff of hot gas. A much larger meteorite, however, may not burn up entirely, so that something of it may hit the Earth's surface. The very largest meteorites can weigh many tons, so that when they hit the Earth's surface they can cause tremendous damage. Luckily, this happens only about once every 100,000 years.

Above: Once in a very long while, a giant meteorite weighing many tons hits the Earth with tremendous force, hollowing out a huge crater. The biggest clearly defined crater of this kind is the Barringer, or Coon Butte, crater in Arizona. It is 574 feet deep and 4,150 feet wide. It was made at least 25,000 years ago. The photograph below was taken from inside, and shows the towering crater wall.

69

The Power of the Stars

When dust and gas from a nebula condense, or concentrate, to form a star, they heat up tremendously until they reach the incredibly high temperature at which stars can shine.

70

THE TEN NEAREST STARS		THE TEN BRIGHTEST STARS			
Name of star	Distance in light-years (One light-year is about 6 trillion miles	Name of star	Type of star	Distance in light-years	What constellation to look at (see also pp. 36–39)
		Sun	small yellow	(8 light-mins.)	—
Proxima Centauri	4.28	Sirius	hot white	8.80	The Great Dog
Alpha Centauri	4.37	Canopus	yellow supergiant	196	The Keel
Alpha Centauri B	4.37	Alpha Centauri	small yellow	4.37	The Centaur
Barnard's star	5.90	Arcturus	giant red	37	The Herdsman
Wolf 359	7.60	Vega	hot white	26	The Lyre
Lalande 21185	8.13	Capella	giant yellow	46	The Charioteer
Sirius	8.80	Rigel	blue giant	815	Orion, The Hunter
Sirius B	8.80	Procyon	small yellow	11.4	The Little Dog
Luyten 726–8 A	8.88	Achenar	hot blue	127	Eridanus
UV Ceti	8.88	Beta Centauri	hot blue	391	The Centaur

Notes: Most of the ten nearest stars are smaller than the Sun. Sirius, though, is both a nearby star and also the brightest star in the sky. Whether a star is bright or not depends not only on how hot or large it is, but how far away it is, as you can see from the right-hand table.

On pages 42–47 we looked in some detail at the nearest and most familiar star, the Sun. Like every other star except the burned-out black dwarf stars (see page 72), the Sun continuously gives out light, heat and other forms of radiation. Poets traditionally have spoken of "the kindly beams of the Sun" because the Sun's radiation is what makes life possible on our planet.

Closer up, however, the Sun or any other active star is seen to be less kindly and more terrifying in its raging energy. A star may go on radiating vast amounts of energy for thousands of millions of years. Where, then, does all this energy come from?

Hydrogen into Helium

The answer is one that people recently experienced for themselves here on Earth when the first hydrogen bomb was exploded in 1952 and destroyed the Bikini atoll in the Pacific ocean. Today's hydrogen bombs are still more powerful, a single one being capable of wiping out completely whole regions of a country.

This gigantic power is exactly the kind possessed by the Sun and other stars, except that they give off their energy continuously and not in a single destructive flash. The awesome power of stars and the hydrogen bomb is produced by a process known as *thermonuclear fusion.*

This is a nuclear reaction in which atomic nuclei fuse together at tremendously high pressures and temperatures. As they fuse, some of the matter, or more accurately, some of the mass of the nuclei is converted into energy. And even a small amount of mass gives rise to a vast amount of energy, which explains the enormous power of a thermonuclear fusion reaction.

In stars, as in a hydrogen bomb, the reaction occurs when hydrogen is converted into helium. The atomic nuclei of two kinds of hydrogen (deuterium and tritium) fuse to form helium nuclei at temperatures of 18,000,000°F or more. Vast amounts of energy are released in the form of heat and light. In a star this reaction continues for as long as there is sufficient hydrogen present.

Starlight twinkles here on Earth not because the star itself varies in the amount of light it puts out, but because of movements in the Earth's own atmosphere. Many stars do vary greatly in the amount of energy they emit, but this variation happens over longer periods of time, such as days, weeks or years.

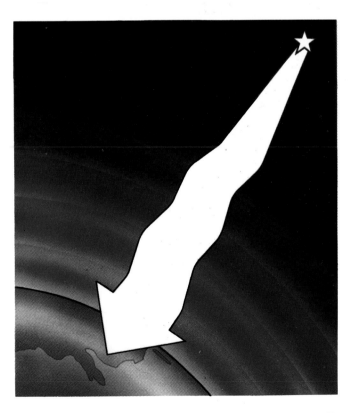

All Kinds of Stars

On the last pages we have taken a look at the lives and histories of stars. Various different kinds of stars were mentioned, which we can now look at separately, together with some further kinds of stars that are new to our picture.

Dwarf Stars

These are stars that are smaller, and generally dimmer, than the Sun. They are of several different kinds. Barnard's star and Proxima Centauri, in the diagram, are red dwarfs, with diameters half or less that of the Sun.

Sirius B, the companion of the brightest of all stars in the sky, Sirius the Dog Star, is a white dwarf. You may remember that this is the kind of star that the Sun will be, near the closing part of its life. Matter in white dwarfs is packed very tightly, so that a teaspoonful would weigh many tons.

Black dwarf stars, also very small, are not shown on the diagram for the simple reason that they do not shine at all and so are almost impossible to find. But they too are worn-out, and probably close-packed, stars.

Neutron Stars or Pulsars

These amazing stars (not shown on the diagram) are far smaller, yet far more dense and heavy, even than a white dwarf star. A neutron star may be only about $12\frac{1}{2}$ miles across, yet weigh as much as the Sun. This means that a single spoonful of it could weigh thousands of tons. Neutron stars are the remnants of the gigantic cosmic explosions called supernovae. They spin at great speed, giving off intense pulses of radiation. For this reason they are called pulsars.

Sunlike Stars

Stars between about three-quarters and twice the mass (weight) of the Sun lead similar lives to that of our own star. On the star diagram Procyon A, for example, is older than the Sun and one and three-quarter times its mass. Procyon A will outgrow the main sequence to become a giant star. Then, much later, it will dwindle to a white dwarf.

Very Hot Stars

Stars that are blue-white in color are hotter than the Sun. They come in several sizes, from about Sunsize to many times the size of the Sun. Sirius, the brightest star in the sky viewed from our particular corner of the Universe, is about two and a quarter times the mass of the Sun. Spica is bigger and hotter still. Rigel is a giant blue-white star, 60,000 times brighter than the Sun and about 600 times brighter than Sirius, but much farther away.

Supergiants

These vast stars are older than blue-white stars; they are what most larger blue-white stars have become with age. On the star diagram, Betelgeuse is a star so large that if the Sun were at its center, Mars would also be inside it. Because it is so spread out, Betelgeuse is thinner in texture, near its surface, than the thinnest gas on Earth. It is much less hot at its surface than the Sun, and so glows with a more reddish color. You can find Betelgeuse easily, as the red star in the left shoulder of the constellation Orion.

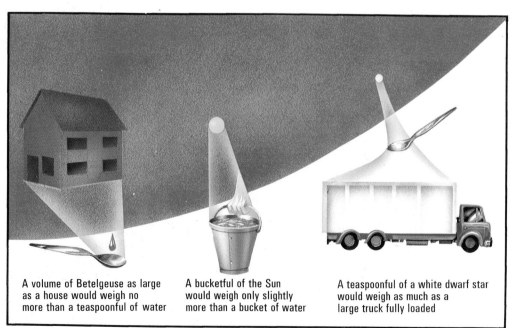

A volume of Betelgeuse as large as a house would weigh no more than a teaspoonful of water

A bucketful of the Sun would weigh only slightly more than a bucket of water

A teaspoonful of a white dwarf star would weigh as much as a large truck fully loaded

Left: The density, or heaviness, of stars varies very greatly. This picture shows the density of three contrasting stars which vary in size from 310,685,500 miles across (Betelgeuse) through about 620,000 miles (the Sun) to only 6,200 miles (a white dwarf). You can also find these stars on the diagram of star types.

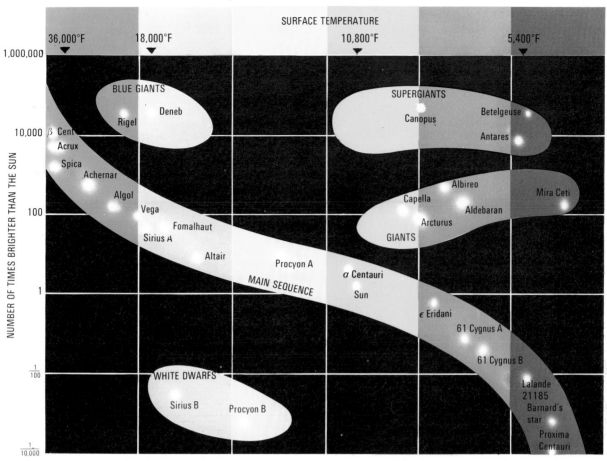

SURFACE TEMPERATURE

36,000°F ▼ 18,000°F ▼ 10,800°F ▼ 5,400°F ▼

1,000,000

BLUE GIANTS
Deneb
Rigel

SUPERGIANTS
Canopus
Betelgeuse
Antares

10,000
β Cent
Acrux
Spica
Achernar
Algol
Vega
Fomalhaut
Sirius A
Altair

Albireo
Capella
Aldebaran
Mira Ceti
Arcturus
GIANTS

100

Procyon A
α Centauri
Sun

MAIN SEQUENCE

1

ε Eridani
61 Cygnus A
61 Cygnus B

1/100

WHITE DWARFS
Sirius B
Procyon B

Lalande
21185
Barnard's
star
Proxima
Centauri

1/10,000

NUMBER OF TIMES BRIGHTER THAN THE SUN

Above: This diagram shows 30 stars grouped according to their size, hotness and brightness. In its lifetime a star may belong in turn to two or more of these groups. For example, our Sun at present is in the main sequence group, but when it gets older it will leave the main sequence and join the group of giant stars. A more massive blue giant star, such as Rigel, will quite soon (in star terms) swell into a supergiant star, then go out with a bang. Only small, rather dim red stars, such as the two called Cygnus, will stay in the main sequence all their very long lives.

Below: Stars are born out of the great masses of dust and gas called nebulae. So astronomers would expect to find many bright, young stars nearby a nebula, such as this one in the constellation Serpens.

Life and Death of Stars

Above: In the constellation Orion, the Hunter, is this brilliant nebula, or great cloud of dust and gas. Its brightness probably results from the birth within it of many thousands of millions of stars, whose combined light makes the whole nebula shine.

Stars are born out of great, swirling clouds of dust and gas in space. Such clouds are found, for example, in the outer spiral arms of our galaxy. In these clouds moving particles will attract one another by the force of gravity. This may cause them to move closer to one another. When this movement gets large enough – that is, when countless millions of particles are moving closer together – a star begins to be born.

The second thing that happens in the birth of a star is that the energy of the moving particles begins to appear as heat. As the particles get ever closer, the amount of heat gets greater and greater, until a glowing ball of gas appears – the young star. As we have seen already on the previous pages, this glowing star energy is produced by a thermonuclear process, which can only start to happen when the temperature of the star rises above about 10 million degrees. What then follows depends very much on the size of the young star.

Star Histories

Stars may all be born more or less in the same way, but they have very different life histories. Longest-lived of stars are those that start out smallest. Compared with other types of stars, these can be thought to live rather dull, if very prolonged, lives. They never become very hot by star standards, even at their brightest glowing only red or yellowish red. During their immense lifetimes, which may stretch over hundreds of thousands of millions of years, these slow-burning stars may swell a little in size and become a little brighter in their prime. But then, as their fuel gives out, they slowly fade away, to become in the end cool, burned-out stars, the nonglowing black dwarfs.

A star of medium size, such as our Sun, has a shorter but more varied life history. Soon after its birth this becomes a very bright star, but then shrinks

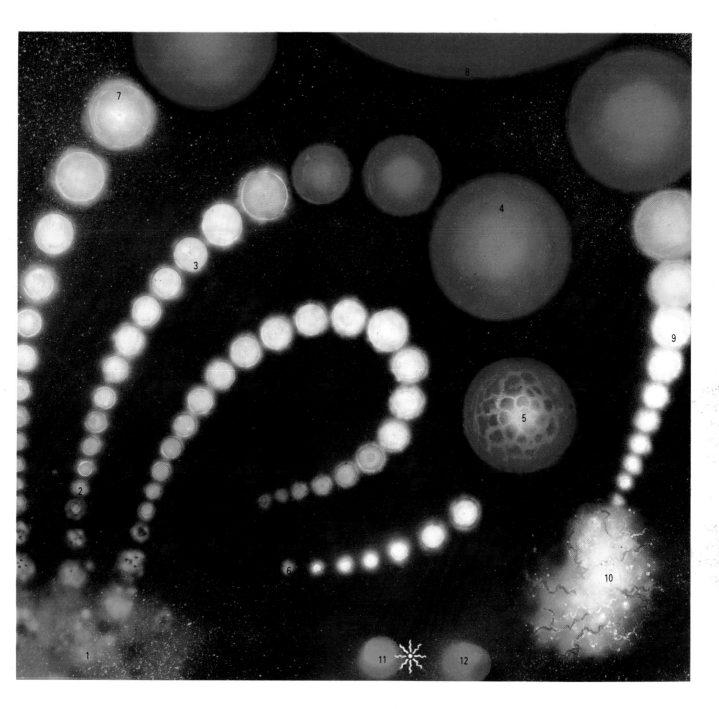

further and cools to become like the Sun is today. As time goes on the fuel at the centre becomes used up. The outer layers begin to burn and the star swells into a red giant. Eventually it becomes too large and unstable and the red giant collapses and shrinks to a white dwarf star, cooler and smaller than the Sun is today. Finally, in a lifetime of about 10,000 million years, the white dwarf cools still further to a black dwarf.

A large young star three or more times the size of the Sun can expect a much briefer and more violent history still, lasting perhaps only a million or so years. Such a star heats up strongly and rapidly, becoming perhaps an intensely hot blue giant. Then, it swells even more greatly to become a supergiant star, still pouring out vast amounts of energy. Such a spendthrift lifestyle cannot last

This diagram shows the different life histories of small, medium-sized and large stars. The numbers show stages in their various life histories. 1 All stars are born from great clouds of dust and gas. 2 This condenses to form the stars. 3 A medium-sized star becomes like our Sun, but 4 then swells into a red giant, then 5 and 6 finally shrinks to a white dwarf. 7 Larger stars grow to be blue giants, then 8 supergiants. 9 The supergiant collapses and explodes, 10 in a supernova. All that may be left is 11 a tiny neutron star, or possibly 12 a black hole. Otherwise, most stars end their days as burned-out black dwarfs.

long. The supergiant quickly uses up most of its nuclear energy, and then it collapses and shrinks, with catastrophic results. Suddenly, a gigantic explosion occurs, flinging most of the star's contents far out into space. All that is left to show where the supergiant once was is a tiny neutron star, or even more strangely, a black hole – which is nothing but gravitational force!

Twins, Clusters and Winkers

Nearly one in three out of all the stars we see in the sky are twins. Most noticeable of all twins is in the constellation Gemini – whose Latin name actually means Twins. Here, the pair of bright stars is Castor and Pollux – you can look them up on the star map on page 36. Other twins are less obvious because one or the other is a faint star, and so you must peer more closely at the night sky to see the pair.

Two stars which look as though they are close together in space may really be so, or they may only look that way because they are in our line of sight, like telegraph poles along a road. In fact, Castor and Pollux are line-of-sight twins, and are really far apart in space. Double stars, or binaries, are genuinely close together. Some binaries can be seen with the naked eye. Clearest of these are the stars Mizar and Alcor, which together make up the middle star in the handle of the Plow. This can be seen all year round in northern skies.

Winkers and Faders

Not all stars shine or twinkle steadily. Many stars vary greatly in brightness, fading for a while and then getting brighter again. Sometimes this is another case of binary stars. One star, darker than the other, circles or orbits around it, so cutting off

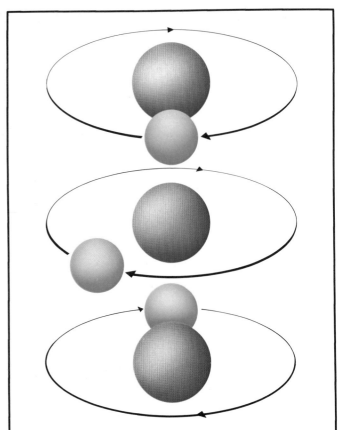

The most famous of the winking stars is Algol, in the constellation Perseus. Astronomers first thought this to be a star that varied very regularly in brightness (every 2 days, 20 hours, 49 minutes), but in 1782 the astronomer John Goodricke put forward a different explanation. He said that the precisely timed variation in brightness of Algol might be due to another body revolving around it, so partly cutting off its light from us at regular intervals. In 1889, with the aid of more powerful modern telescopes, this was shown to be true. The twin star that partly cuts off Algol's light is actually the bigger of the two, but it is also a much dimmer star.

its light regularly. The first of these winking binaries to be discovered is called Algol. It is shown in the diagram above.

In other cases, a star really does decrease and increase in brightness. This happens with many stars called Cepheid Variables – Cepheid because they were first noticed in the constellation Cepheus, and Variable because they fade and brighten with great regularity. Many other Cepheid Variables have been discovered. The Pole Star, Polaris, is one that varies in brightness over a period of just under four days.

Other variable stars vary in brightness much less regularly. Some of these irregular variables are

This famous cluster of stars, just visible to the naked eye, is in the constellation Hercules. It is numbered M13 and consists of some tens of thousands of stars regularly arranged in a ball-shaped mass. For this reason M13 is known as a globular cluster. Its stars are yellow in color and are thought to be very old.

giant stars that behave in an unstable way, being sometimes larger and brighter, at other times less large and dimmer. Easiest to find is Betelgeuse, the red supergiant in the left shoulder of Orion. But still more remarkable in this way is Mira, another great giant star, in the constellation Cetus. Mira varies so much in brightness that at some times it is a bright star and at others it is too dim to be seen without a telescope.

Yet other stars flare up just once, then return to something like their former brightness, or disappear from sight for ever. In our own galaxy, about 25 stars flare up each year. Such a flare-up is called by the name nova, unless it is extra bright and the star disappears. Then, it is one of the gigantic star explosions called a supernova.

Clusters

In our own part of the Universe, several stars are fairly close to one another, so that our Sun has a number of neighbors. Among them are two small stars in the constellation Centaurus (see the table on page 71). In many other, more crowded parts of

the Universe, greater numbers of stars are grouped together in clusters.

Star clusters are mainly of two different kinds. One kind is the open cluster, which shows up as an irregular patch of starry light in the sky. Most famously visible of open clusters is the patch called the Pleiades. Both this and a smaller patch, called the Hyades, are in the constellation Taurus, the Bull.

The other main kind of cluster is the globular (rounded) cluster. Open clusters contain up to thousands of members, globular clusters contain up to 10 million stars, and their stars are so crowded that most are difficult to tell apart.

Nebulae

Stars, as we have seen, are born in nebulae. These are great clouds of dust and gas, scattered widely throughout the Universe. The shapes of some nebulae are similar to recognizable objects. This has led to such names as the Horsehead nebula and the Coal Sack nebula. Most nebulae, however, appear through the telescope as formless clouds in space. The name *nebula* comes from the Latin word for cloud.

Dust and gas, the substances of nebulae, are really spread throughout the Universe. All the same, the vast tracts of empty space are well named because on average they only contain one atom of matter for every cubic centimeter (6/100 inch). The dust and gas in nebulae is much more concentrated than in interstellar (between-stars) space, though it is still of course spread much thinner than in our own air.

For this reason, nebulae are found where stars are also found, namely in the great islands of stars known as the galaxies. Those of our own galaxy are called *galactic nebulae*, and those belonging to other galaxies are called *extra-galactic nebulae*.

Nebulae may be dark clouds, like the Horsehead and Coal Sack nebulae, or they may be shining clouds. There are two main kinds of shining nebulae. In some, the dust and gas shine rather faintly by light reflected from the stars seen within them. In others, the stars are so hot that they cause the whole nebula to glow brightly.

Nebulae include not only the birthplaces of stars but also their graveyards. Perhaps the planetary nebulae shown in the small pictures are really old giant stars puffing off their outer layers into space. But the most dramatic deathplaces of stars are those other shining nebulae, such as the Crab nebula, that have resulted from supernovae, where stars have blown up in the most catastrophic way.

Right: One of the most famous of the dark nebulae, the Horsehead nebula. This is clearly named after its shape, which is rather like that of the head of a chess knight. With the aid of a small telescope, or field glasses, you can find this dark nebula in the constellation of Orion, the Hunter – just below his belt of stars.

Inset, top: The Lagoon nebula in the constellation Sagittarius is an entirely different object from a dark nebula, which is unlit dust and gas. The Lagoon nebula is made to glow by the radiation from a nearby bright star. Its dark "lagoon" is caused by dark, obscuring matter in the foreground.

Inset, bottom: The Veil nebula in the constellation Cygnus. It glows as it collides with the dust and gas of interstellar space. The blue light comes from the leading edge of the nebula, where the collisions are hardest.

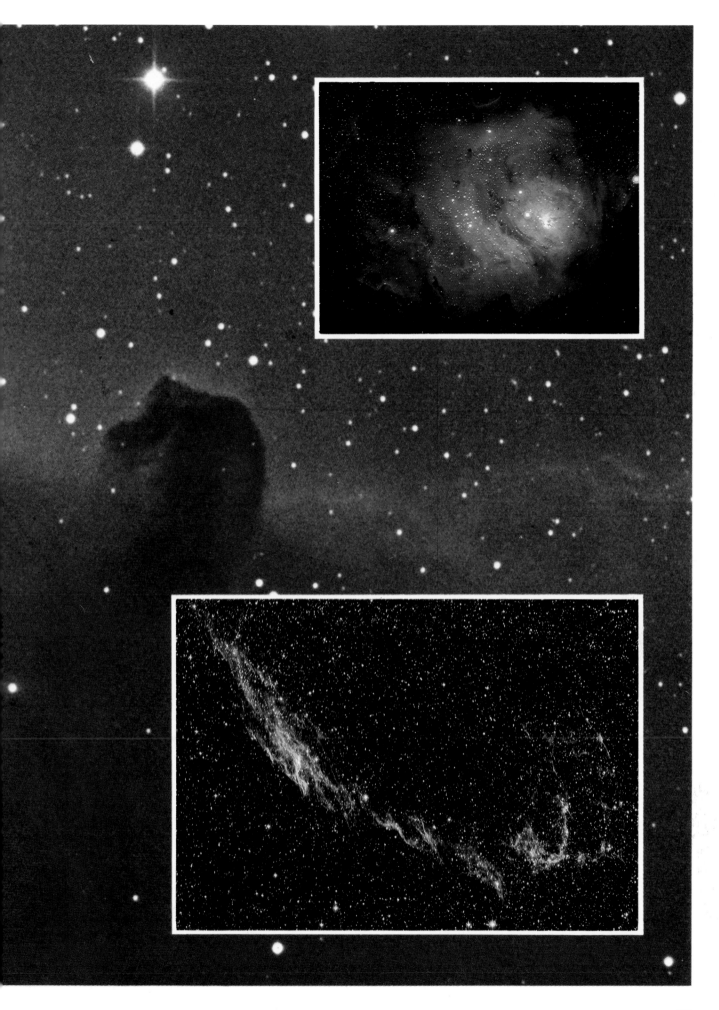

The Milky Way

A view of one star-crowded part of the Milky Way. This photograph was taken through a telescope pointing in the direction of the constellation Cygnus, the Swan. As you can see, many stars are hidden from view by space dust. When you look at the Milky Way, you are looking at the thickest part of our own galaxy; most of its stars are so faint, or so hidden by dust and nebulae, that we see only their combined glow in the sky.

Look up into the sky on a clear, moonless night and you will see an irregular, hazy band of light crossing the heavens. This is called the Milky Way, and is part of our own galaxy – a vast island population of stars of which the Sun is just one rather insignificant member.

The Shape of the Galaxy

Our galaxy is shaped rather like a fried egg – a disk with a thicker middle and a thinner rim or edge. Of course, we can never see this shape because the Earth and Solar System are inside the disk. When you look at the Milky Way, you are really gazing into the thickest part of the Galaxy, towards the center. When you look at other parts of the sky, you are looking away from the Galaxy, out into empty space.

Most stars of the Milky Way are so far and faint, or so obscured by space dust and nebulae that come between the Earth and themselves, that we do not see them individually. We can only see their combined glow, the faintly shining band of the Milky Way.

The number of stars is not equal in all parts of the Galaxy. In some parts stars are quite rare; in others, such as our own, stars are scattered about rather averagely. In yet other parts, called star clouds, stars are packed much more closely together. One such star cloud can be seen in the constellation Sagittarius, the Archer. This star cloud is in the thicker "egg yolk" part of the Galaxy, so that when you look in the direction of the Archer, you are looking into the center of the Galaxy.

The Spinning Galaxy
If we could see it from far away in space, the Milky Way would look rather like a pinwheel firework, because from its rim come a number of starry spiral arms like those of the lighted, spinning firework. Our galaxy also seems to have got its spiral-armed shape by a spinning process. This process, though, has lasted longer than the firework, for our galaxy is about 12,000 million years old!

The Spiral's Size and Speed
Our Solar System lies fairly far out along one of the spiral arms of the Milky Way. It is about 20,000 light-years from the rim of the galaxy and about 30,000 light-years from its center. The Galaxy itself is about 100,000 light-years across at its widest. (A light-year is the enormous distance of about 5.88 trillion miles.) At its center, the Galaxy is 13,000 light-years deep, whereas at its outer edges it is more like 5,000 light-years deep. The whole galaxy contains more than 100,000 million stars.

Our galaxy carries the Solar System with it in its spinning movement, so that the Sun and its planets travel in a series of great loops or circuits within the Galaxy. Each of these circuits takes 225 million years, and the speed of our Solar System in the circuit is about 155 miles per second. Nearer the center of the Galaxy, speeds of rotation are greater, and farther out towards the edge of the galaxy, slower. This in itself is rather unexpected, when you consider that the rim of a wheel always rotates faster than the hub at its center.

Above: An artist's view of our galaxy, seen from the side. The arrow shows the position of our Sun and its planets. Of course, no one has ever witnessed this view because no astronaut has been anywhere near the edge of our galaxy. But there are many millions of other galaxies in the Universe like our own. One of these is shown below. It is a spiral galaxy in the constellation Andromeda. This constellation also contains the nearest of all other galaxies to our own, the Andromeda spiral galaxy.

Below: An artist's view of our galaxy, showing its spiral, pinwheel shape. Again, the arrow points to the position of our Solar System. This shows why, when we look towards the center of our galaxy, we see more stars than when we look out towards its edge.

Other Galaxies

A galaxy is a great island of stars and their planets, nebulae and interstellar dust and gas, turning in the loneliness of empty space. Astronomers have now recorded a very great number of other galaxies beyond our own, many of them so far away that despite their gigantic size, the most powerful of telescopes is needed to show them.

Some of the other galaxies look very like our own galaxy, that is, they are disk shaped with many arms spiralling out from the disk. Other galaxies have a rather similar shape, but the spiral arms are longer or shorter, the central hub bigger or smaller, as shown in the pictures below. Sometimes the arms may spiral right around to make the "barred spiral" shape. Yet other galaxies are not spiral at all but are more globular or ball shaped, or are completely irregular in shape, like a giant nebula.

Galaxies, like the stars they contain, are more common in some parts of space than in others. The most powerful modern telescopes reveal that throughout the Universe, galaxies usually occur in groups or clusters, with much larger volumes of empty space between. Our own galaxy, together with its nearest neighbor, the great spiral Andromeda galaxy, belongs to a cluster of about 20 galaxies, which we know as the Local Group. Altogether, this Local Group of galaxies occupies a space about five million light-years across.

Far beyond our Local Group are immense numbers of other galaxies, stretching away to the limits of the known Universe. And strangely, the farther away they are, the faster they are moving away from us.

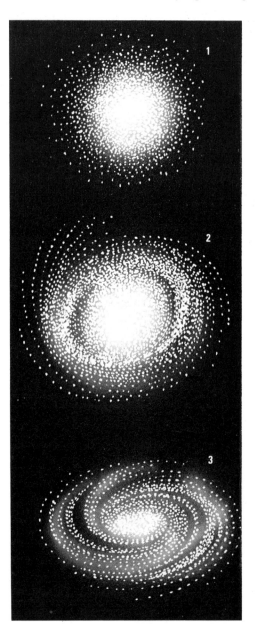

The numbered pictures show various kinds of galaxies. (1) is an elliptical galaxy, with a rather ball-like shape. Some elliptical galaxies look more squashed together than this. Elliptical galaxies contain mainly old stars, as do the central hubs of the spiral galaxies. (2) is a spiral galaxy with a large hub and rather short spiral arms. The spiral arms of galaxies generally contain younger stars. (3) and (4) are more open types of spiral galaxy, similar to our own galaxy and the great galaxy in Andromeda. (5) is a barred-spiral galaxy, in which the spiral arms surround the hub completely. (6) is an irregular galaxy, shaped more like a ragged nebula than like the other, more symmetrical galaxies.

Opposite, inset above: A spiral galaxy in the constellation Triangulum, with widely separated arms.

Opposite: The Andromeda galaxy, the nearest spiral to our own Milky Way galaxy, is probably similar in size. This is how our galaxy would appear if we could see it from out in space.

Opposite, inset below: Another spiral galaxy in the constellation Sculptor. As galaxies go, this one is fairly near to us – a mere 8 million light-years away. Unlike the vast majority of other galaxies, it appears to be moving towards us.

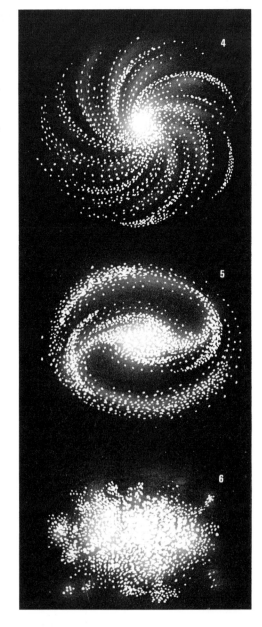

Rockets into Space

As fireworks, rockets are very familiar. Nearly everyone has seen them shooting up into the sky. Firework rockets were invented by the Chinese, about a thousand years ago. The thrust that sends a firework rocket up into the sky is provided by fast-burning gunpowder. This makes a great deal of gas and smoke, which squirts out from the rocket to push it rapidly through the air.

Other, larger, kinds of rockets also work on this principle of thrust by a burning fuel, although usually they do not burn solid gunpowder as their fuel. Rocket cars and rocket trains, for example, have been designed to burn kerosene or other high-energy liquid fuels to provide the fierce thrust that sends them across land at great speeds.

The First Rockets

Nearly all the basic ideas of modern space rockets were first thought up about 80 years ago, by a Russian schoolteacher named Konstantin Tsiolkovsky. Some people said that a rocket could not travel in empty space, because there was no air to push against. But Tsiolkovsky argued, quite correctly, that a rocket would push against its own gases and so thrust itself through space.

Tsiolkovsky also suggested that one or more large rockets should be used to carry a small rocket far up into the atmosphere before the small rocket was fired to carry on into outer space. He also recommended liquid fuels for the rocket thrust, and not the traditional solid ones such as gunpowder.

Despite all his brilliant ideas, Tsiolkovsky remained rather little known, and it was in the United States of America, rather than in Russia, that the first large, liquid-fueled rocket was fired off successfully in 1926. This was the invention of Robert H. Goddard, a rocket expert who had also written about the possibility of space travel. His rocket did not reach up anywhere as far as outer space, but all the same, it was the true ancestor of modern giant space rockets. Like them, it got its thrust from the mixture of a liquid fuel, such as gasoline and kerosene, with liquid oxygen. When these liquids are mixed and ignited, an intensely powerful combustion, or burning, results, and the gases from this burning drive the rocket up into the air.

In the Second World War German engineers invented liquid-fueled flying weapons that traveled 124 miles or more through the air before diving down to create a damaging explosion. The first of these flying bombs, the V1 or "doodlebug," was really an early, pilotless jet aircraft. Jets, like rockets, travel through the air by the thrust of gases from their burning fuel, although unlike rockets, they need a supply of air to do so. The second type of German flying bomb was called the V2. This flew higher and faster than the V1, and delivered a more devastating explosion. The V2 was a true rocket because it did not need to suck in air to burn its fuel. It was the immediate ancestor of modern space rockets.

After the Second World War, the big German V2 war rocket was used in the United States and USSR as the starting model for space rocket design. The V2 was a single stage rocket, but the later space rockets were in three stages, with a small rocket on top carried up by two larger rockets below. The space age began on October 4, 1957, when a three-stage rocket rose from the USSR to put the first artificial satellite into space orbit around the Earth.

Below: Large rockets of today are liquid fueled. The fuel, usually kerosene, is stored in a tank inside the rocket. A second tank contains liquid oxygen. To fire the rocket, these two chemical substances are brought together by means of various pumps and mixers. This happens at the tail end of the rocket and results in an intensely fierce burning, or combustion. Hot gases blast out from the exhaust nozzle of the rocket to drive the rocket upwards.

LIQUID PROPELLANT ROCKET

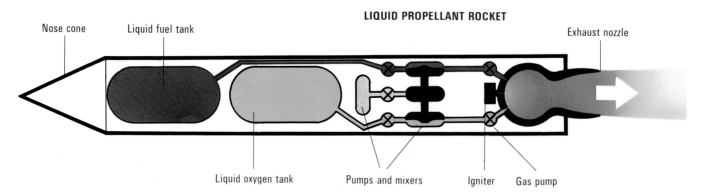

Nose cone   Liquid fuel tank   Exhaust nozzle

Liquid oxygen tank   Pumps and mixers   Igniter   Gas pump

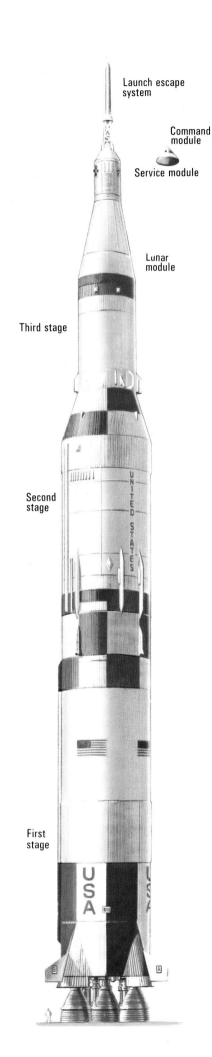

Launch escape
system

Command
module

Service module

Lunar
module

Third stage

Second
stage

First
stage

Above: In July 1969 a Saturn V rocket blasts off to start the Apollo 11 crew, in their small spacecraft, on their epic journey to become the first men on the Moon.

Left: Giant space rockets, such as this Saturn moon rocket, usually consist of three rockets mounted one on top of the other. This arrangement is called a *three-stage rocket*. The two larger, lower rockets carry the small, uppermost rocket far up into the atmosphere. There they drop off, in turn, and the smallest rocket fires to propel the spacecraft out into space and, in this case, as far as the Moon. The spacecraft itself has further small rockets, with which it is able to get back from the Moon to Earth.

The First Satellites

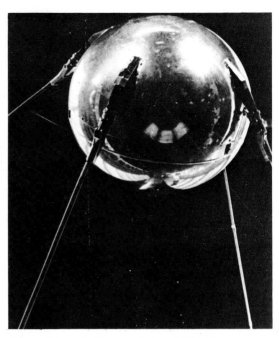

By the early 1950s a number of large rockets, developed from the German V2 war rocket, had penetrated the Earth's atmosphere to reach empty space. But they lacked enough power to allow them to escape from Earth's gravity or even to take them into orbit around the Earth. So in all cases, they fell back again to Earth after their lofty flights.

On October 4, 1957, the Earth's first artificial satellite rose into orbit. Circling the globe every 96 minutes, Sputnik 1, as it was called, regularly uttered a bleeping signal which became instantly famous on the world's radio. The flight of Sputnik 1 was the achievement of Russian rocket engineers, but the equally advanced Americans were not long behind with their first artificial satellite, Explorer 1, which they placed in orbit on January 31, 1958.

After this breakthrough, satellites from both countries followed one another thick and fast into space, in the beginnings of what came to be called the *space race*. Satellites soon became larger, heavier and more elaborate. Sputnik 1 was only 1½ feet or so across and weighed about 185 pounds. A year or so later, both the USSR and USA had put up satellites of a ton or more in weight.

The bigger satellites did more than bleep. Even before the first American satellite left for space, the Russians had put up the first living creature from Earth to enter space – a dog named Laika inside a half-ton orbiter. More space-dogs followed from the USSR, but everyone was waiting for the event for which these were just a preparation – the first human spaceflight. This was also to be a Russian achievement, but meanwhile, other rapid technical advances were being made in the satellites of both countries.

Sophisticated Satellites

The very first satellites went up into space simply to conquer Earth's gravity, as Isaac Newton had said was possible 250 years previously (see page 20). Soon, however, satellites were being put up to do special scientific jobs lasting months or even years. They needed a regular supply of electric power for this. To meet this need came one of the first major developments in satellite technology, *solar panels*, which make electricity directly from sunlight. You are probably familiar with the look of these because they appear on nearly every picture of a satellite.

Among other things, these solar panels gave a satellite enough power to send back radio messages and television pictures to Earth. For the first time, people on Earth were able to see what their world looked like from outer space. Some of these early satellites, fitted with television cameras, sent back many thousands of detailed pictures of the clouds

Above: First of all man-made satellites in space was Russia's Sputnik 1. Its silver ball measured about 14 inches across, but this was increased to about 1½ feet by its four long antennas. Bigger satellites soon followed, both from Russia and from America.

Below: Among the first satellites designed specially for weather forecasting was Tiros 1, launched in April 1960. It radioed detailed pictures of clouds back to Earth. This enabled American weather men to give longer warnings of violent storms, such as the typhoons that regularly attack the south-eastern coastline of the United States.

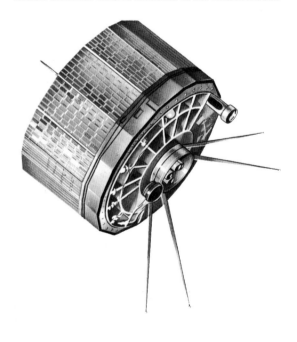

Right: Yuri Gagarin was the first Russian astronaut and also the first man in space. He survived his one-orbit trip around the world safely enough, but died in an airplane crash in 1968.

VOSTOK

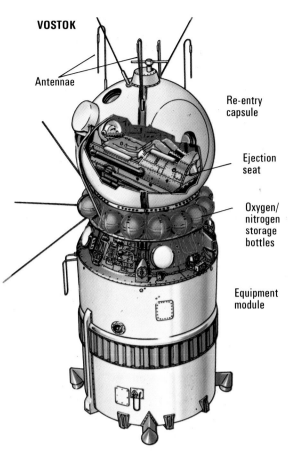

Antennae

Re-entry capsule

Ejection seat

Oxygen/ nitrogen storage bottles

Equipment module

Above: Vostok 1, the spacecraft in which Yuri Gagarin made the first orbital flight around Earth on April 12, 1961. You can see the astronaut lying in his spherical capsule, which was the only part of the spacecraft he occupied and in which he returned to Earth.

covering many parts of the globe. Never before had it been possible to see so far, to tell what kind of weather was coming.

These early satellites also made it possible to send messages more easily between faraway places on Earth. The first kinds of *communication satellites* worked as simple reflectors in space for radio waves sent up from broadcasting stations on the Earth's surface. When the radio waves bounced off the satellite, they could often be received as far away as the other side of the world.

Yet other satellites of the first three years of the space age were used for what seemed like more specialized scientific tasks. Some measured the intensity of particular kinds of radiation reaching Earth from space, such as ultraviolet rays from the Sun, X-rays and cosmic rays. This kind of measurement had a very practical use for judging the safety of space flights for human beings, who might be harmed or even killed by the rays. Also useful for the same reason were satellite measurements of the number of tiny meteorites hurtling through space – the dust-sized bits of rock and metal that ended by burning up in Earth's atmosphere.

Human Beings Reach Space

In the first few months of the space age, Russia sent up dogs into space and later recovered some of these animals. They looked none the worse for their ordeal, despite space radiation and meteorites. Nearly two and a half years later, on April 12, 1961, a rocket rose from Russia with the first spaceman, or astronaut, on board. His name was Yuri Gagarin. Some 108 minutes later, after completing just one orbit of the Earth in his spacecraft, Vostok 1, at heights between 105 and 195 miles, he returned safely to the Russian land surface, floating down in his round metal capsule by parachute.

It took the American space experts nearly a year to catch up with their Russian rivals. In 1961 two of their rockets carried a man in *sub-orbital* flight, that is, high up and part of the way, but not all the way, around the Earth. But it was not until early in the next year, on February 20, 1962, that an American rose into orbit. His name is John Glenn and he orbited our world three times in his spacecraft Friendship 7, which was of the Mercury type shown in the picture.

Below: This Mercury spacecraft is of the kind that carried the first American astronaut, John Glenn, around the Earth on February 20, 1962. After his three-orbit flight, Glenn, like Gagarin almost a year earlier, floated safely back to Earth by parachute, inside his metal capsule.

MERCURY

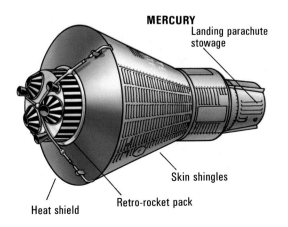

Landing parachute stowage

Skin shingles

Retro-rocket pack

Heat shield

Men into Orbit

The first ten years of the space age saw dozens of manned space missions, as they came to be called, from both the United States and the Soviet Union. This was the time when men first learned to live and work well away from their home planet, in the strangely unfamiliar conditions of airlessness and weightlessness. The number of orbital flights made by an astronaut quickly rose from one or a few to twenty or more. In another few years, an astronaut might remain in orbit for as many as a hundred trips around his native globe.

It was at this time, in the mid 1960s, that spacecraft began to carry two or even three astronauts up into orbit. This left one or more astronauts free to carry out experiments, while the remaining crew member piloted the spacecraft.

The next great advance came when orbiting astronauts began to leave their spacecraft for what were called space walks. First of the space walkers was the Russian Alexei Leonov, who in March 1965 climbed out from his two-man orbiter Voskhod 2 and floundered and floated weightless in empty space for about 10 minutes. The first American space walker was Ed White, who spent 20 minutes outside his Gemini spacecraft in June 1965. Like later space walkers, both men were tethered safely to their orbiting craft by a strong lifeline, which allowed them to move about in space beside their craft without the risk of floating off forever.

The pilots who stayed behind in the spacecraft sometimes had the more difficult jobs. In the first space walker's own craft, Voskhod 2, the pilot was Pavel Belyayev. When the time came after 17 orbits to dip down through the Earth's atmosphere, the automatic guider and lander device of the spacecraft would not work properly. This meant that Belyayev had to guide his craft to Earth, after making one more orbit than had been planned. And this, in turn, meant that he could only land Voskhod 2 at a place far distant from the chosen spot – in fact, way out in the frozen Siberian forest. However, Voskhod 2 was soon recovered, with both its astronauts unharmed. In general, spaceflight was proving not as dangerous as feared, although fatal accidents did sometimes occur. Three American astronauts were killed by fire in their Apollo spacecraft in 1967, while it was still on the ground, and in 1971, three Russian astronauts were killed while returning to Earth.

An American spacecraft floats in the calm waters of the Pacific Ocean, while Navy frogmen undo the hatches to release the astronauts after their splashdown. Russian spacecraft, by contrast, touch down to solid ground. But in both cases, the final drop of about 6 miles is made with the support of parachutes, which slow the spacecraft down after their swift, intensely hot trip through the Earth's atmosphere.

Right: The first American astronaut to climb outside his orbiting spacecraft was Edward H. White. His space walk took place in June 1965, a few months after the Russian Alexei Leonov made the first space walk of all. These astronauts were not really as unsafe as they looked, poised a hundred or more miles in space above the glowing Earth. You can see the lifeline that holds astronaut White to his Gemini spacecraft. This not only prevents him from floating away into space, but also supplies him with oxygen vital to his breathing. His thick, padded spacesuit protects, or insulates, him from both the heat and cold of space. It also acts as a cushion against injury from any small meteorite that should happen along.

GEMINI

Right: The Gemini spacecraft, the type from which Edward White made his space walk, were among the pioneering craft of the early space age. Carried up into space by Titan rockets, they made altogether many hundreds of journeys around the Earth. It was these spacecraft, too, that were first brought together in space in what was called a *space rendezvous*. This happened with Gemini 6 and Gemini 7 in December, 1965. After its mission was completed, a Gemini, like other kinds or orbiting spacecraft, was started on its pathway back to Earth by the firing of small rockets in the direction of its orbit. This slowed the spacecraft down so that it took a lower and lower pathway until it entered the Earth's atmosphere at about 17,400 miles per hour. At this tremendous speed, the spacecraft's heat shield took most of the air friction. It glowed red-hot but protected the rest of the spacecraft from overheating by the violent rush of air.

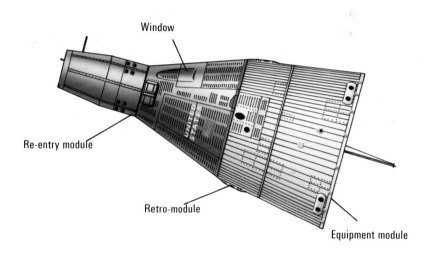

Window

Re-entry module

Retro-module

Equipment module

SOYUZ

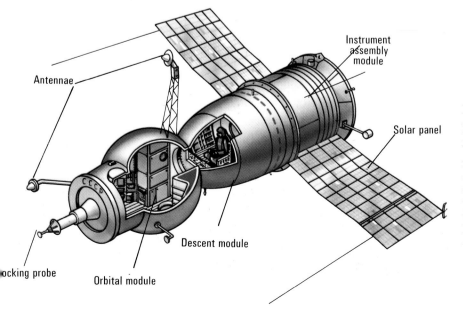

Instrument assembly module

Antennae

Solar panel

Descent module

ocking probe

Orbital module

Left: The Soyuz spacecraft carried two or three astronauts and was used to supply crews for the Russian Salyut space stations in the early 1970's. These space missions were generally very successful, and the USSR has since concentrated on the orbiting station or laboratory for many of its space achievements. But the Soyuz spacecraft was also the scene of one of the few space tragedies, when three Russian astronauts were killed while returning to Earth in Soyuz 11. An equal previous tragedy had been the death of three American astronauts by fire in their Apollo spacecraft, while it was still on the ground.

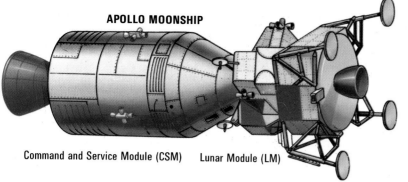

APOLLO MOONSHIP

Command and Service Module (CSM)   Lunar Module (LM)

Left: The complete Apollo moonship, as taken up into space by the giant Saturn rocket. Before each Moon landing the moonship divided, or undocked, into two main parts or modules, called for short the CSM and the LM.

Apollo Beginnings

The Apollo Project was the first to land human beings on another world – Earth's satellite and nearest neighbor, the Moon. In the space race between the United States and the Soviet Union, Apollo put the United States into an unchallenged lead.

Altogether there were 17 Apollo flights, or missions, between the years 1966 and 1972. The first six of these missions, Apollos 1 to 6, were test flights in which the giant Saturn rockets carried up Apollo spacecraft which had no astronauts on board. Apollos 7 to 10 carried astronauts and equipment into space and as far as Moon orbit, 248,548 miles away from Earth.

APOLLO

Command module

Rocket fuel tanks

Pressure

Service module engine nozzle

Jets for steering mc

Fuel cells to provide electricity

Service module

Above: The Command and Service Module, or CSM. This consists of two main parts: the Command Module, which contains crew and controls, and the Service Module, which contains the space rocket and other power units. The crew cabin is filled with air at normal Earth pressure, allowing astronauts to breathe freely. The pointed nose of the CSM contains a hatch or air lock by which two members of the crew enter the LM, before it undocks for a trip down to the Moon. The third crew member remains in the CSM in Moon orbit.

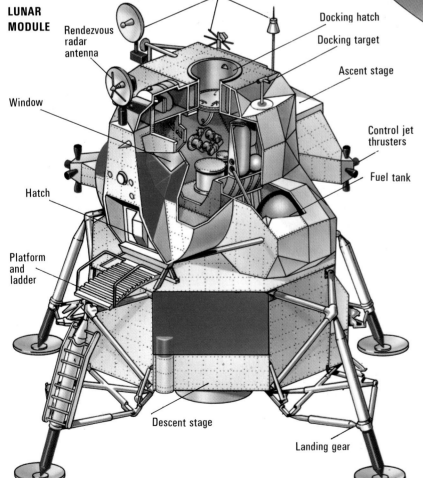

LUNAR MODULE

Radio antennae

Rendezvous radar antenna

Docking hatch

Docking target

Ascent stage

Window

Control jet thrusters

Fuel tank

Hatch

Platform and ladder

Descent stage

Landing gear

Left: The Lunar Module, or LM. This also consists of two main parts. The lower part is called the Descent Stage because it contains the rocket engine which acts as a brake when the LM touches down on the Moon. The upper part of the LM is called the Ascent Stage. Only this stage rises from the Moon to carry the two Moon explorers back into orbit, there to dock with the CSM. The Descent Stage is left behind on the Moon's surface after a slight rocket push lifts the Ascent Stage against the Moon's weak gravity. Then, after the two explorers have docked and transferred back into the crew cabin, the Ascent Stage too is allowed to go, to crash on to the Moon's surface. Finally, the CSM's rocket motor is fired to take it and the three astronauts out of Moon orbit and back towards Earth. The complete Apollo journey to the Moon and back is shown on the right.

Before any Moon landing could be attempted, many technical tests had to be carried out in space. Most important of these were the "rendezvous and docking" tests. This means the dividing and rejoining of the two main parts of the Apollo spacecraft, or moonship. It was vitally necessary to get this right, because when the first Moon landings were made, one part of the moonship (CSM) would stay in orbit around the Moon, while the other part (LM) would separate off, land on the Moon, then rejoin the CSM in orbit (rendezvous and docking) for the journey back to Earth. The pictures show all the main equipment needed for these early Apollo missions, and details of the journey to and from Moon orbit.

Above left: An Apollo 9 astronaut, Dave Scott, stands in the open hatch of the Command Module, while his spacecraft is in orbit. Behind him you see the curved surface of Earth. This Apollo mission was the one which tested "rendezvous and docking" in Earth orbit before the Moon flights.

Above: The full space suit worn by Apollo astronauts whenever they left their spacecraft, either in space or on the Moon. (Dave Scott is wearing one, above left.) The space suit allows an astronaut to move about unharmed in hostile space conditions. It insulates, or protects, the wearer equally against the killing cold and heat of space. It cushions him against knocks and even blows from small meteorites. It provides him with vital oxygen, and removes the waste products from his body into a bag, for later disposal.

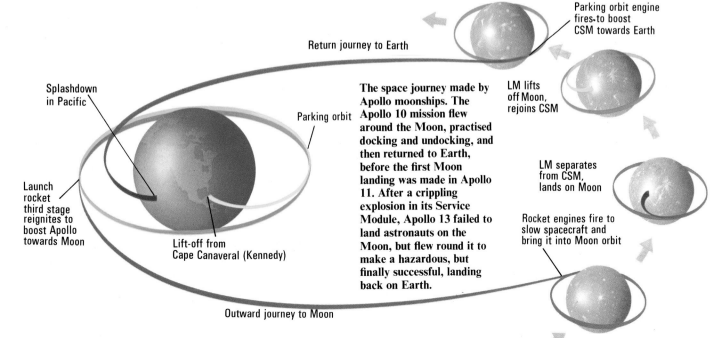

Return journey to Earth

Parking orbit engine fires to boost CSM towards Earth

LM lifts off Moon, rejoins CSM

Splashdown in Pacific

Parking orbit

Launch rocket third stage reignites to boost Apollo towards Moon

Lift-off from Cape Canaveral (Kennedy)

LM separates from CSM, lands on Moon

Rocket engines fire to slow spacecraft and bring it into Moon orbit

The space journey made by Apollo moonships. The Apollo 10 mission flew around the Moon, practised docking and undocking, and then returned to Earth, before the first Moon landing was made in Apollo 11. After a crippling explosion in its Service Module, Apollo 13 failed to land astronauts on the Moon, but flew round it to make a hazardous, but finally successful, landing back on Earth.

Outward journey to Moon

Men on the Moon

On July 20, 1969, the Lunar Module *Eagle* undocked from the Command Module *Columbia* and began its short descent to the surface of the Moon. Not long afterwards, back on Earth, the biggest television audience ever assembled saw the first Moon landing, and for the first time a human put his foot upon an alien world. This "giant step for Mankind," as Neil Armstrong, the explorer, called it, instantly became world famous.

The first Moon astronauts, Armstrong and Edwin Aldrin, were followed by 10 more Apollo crew members who walked, rode and worked on the Moon's surface between 1969 and 1972. They explored, in all, some hundreds of square miles of the Moon's surface. Between them they collected several hundred pounds of Moon rock for geologists back on Earth to study. They set up many scientific instruments on the Moon, some of which continue to work even now.

Among the most important of these instruments are the seismographs, which record moonquakes similar to the earthquakes experienced in our own world. Over the years, these seismographs have shown that although the Moon is a dead world on its surface, down below it is still active to some extent, because Moon tremors, or small moonquakes, happen fairly frequently.

Human explorers said goodbye to the Moon, probably for this century at least, when Eugene Cernan and Harrison Schmitt lifted from its surface in December 1972. But by this time, dozens of manned and unmanned spacecraft from Earth had visited our dusty satellite.

Above: The first people to step on to the Moon's surface were Neil Armstrong and Edwin Aldrin of the Apollo 11 mission. Both of them are visible in this picture; you can see Armstrong, the mission commander and the first man on the Moon, reflected in the visor of Aldrin's space helmet. Also reflected is the Lunar Module, or LM, that brought the first two Moon explorers to the surface of that airless world.

Below: Not every piece of equipment carried on board Apollo spacecraft was super-scientific. This picture shows a piece of home-made equipment, used to get rid of carbon dioxide gas from the spacecraft's atmosphere. It was an emergency invention of the Apollo 13 astronauts, to help them to get back home safely after an explosion in their Service Module or main rocket unit.

Below: This remarkable photograph was taken from the Apollo 11 Command Module, orbiting the Moon in July 1969, and shows the Lunar Module not far away. Over the bare, pitted surface of the Moon rises a bright, distant Earth.

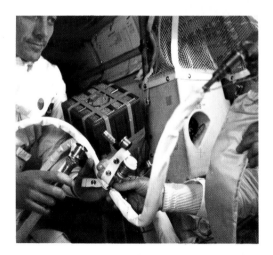

Above: This picture is a scene from the last of the Apollo missions, Apollo 17. It shows Harrison Schmitt, who is both an astronaut and a geologist. He is collecting samples of lunar soil with a special rake. Notice his large back-pack, containing life-support systems. Notice also the desolate, brightly lit Moon surface compared with the dark lunar daytime sky. The lunar sky is not lit up, like an Earth daytime sky, because there is no air to scatter the sunlight.

Above: Another scene from the last of the Apollo Moon explorations. Here, the Apollo 17 commander, Eugene Cernan, is driving the Lunar Rover or "moon buggy." In this buggy the astronauts, together with their collecting equipment, roamed the Moon's surface for a total of 22 miles, collecting more than 250 pounds of lunar rock samples which they took back to Earth.

Right: Harrison Schmitt and the moon buggy again. The mission geologist is poised by a huge lunar rock, while collecting samples. The umbrellalike object at the back of the Lunar Rover is a radio antenna. A moon buggy was carried by the last three Apollo missions. Schmitt and Cernan stayed on the Moon for 22 hours, longer than any of the previous Apollo astronauts.

Moon Machines

The Apollo project sent men to the Moon, but exploration of the Moon did not begin, nor quite end, with the Apollo astronauts. For 10 years before the first man stepped on the Moon, unmanned Luna spacecraft from the USSR had been aimed at the Moon, and the USA had followed by sending unmanned Ranger probes.

The earliest of these probes had simply crashed into the Moon's surface, usually shortly after taking and sending back close-up pictures of the Moon's surface, or after throwing out a package of instruments which then took pictures. Later probes spun around the Moon and took close-up photographs at leisure. The first of these was the Russian Luna 10, which orbited the Moon in March 1966. American Lunar Orbiter probes made a detailed survey of the Moon from August

1966 to late 1967. These spacecraft were, of course, controlled by human controllers back on Earth, who with the aid of their computers sent out radio instructions for each new move that the spacecraft or their instruments made.

"Soft" Landings

The first spacecraft to make a safe, "soft" landing on the Moon was Luna 9, sent from Russia in January 1966. This was soon followed by Surveyor 1 from the United States, which soft-landed in May of that year. The successful landing of these mooncraft proved, among other things, that the Moon was not entirely covered in deep dust that would swallow up any vehicle sent there – an important point to be sure of before the first people set out for the Moon.

These and other early Moon machines made a wide-ranging survey of the Moon, which provided Earth scientists with a detailed technical picture of our neighboring world some years before the first Moon astronauts set up their instruments there.

Left: An assortment of strange Moon machines, sent up over a period of 66 years between 1964 and 1970. Of course, they were never all together like this.

In the lunar sky are, left, Ranger 7 (USA, 1964) which took thousands of pictures before it crash-landed; and right, Luna 12 (USSR, 1966), which was the first probe to send television pictures back from lunar orbit.

On the lunar surface are four Moon machines. Left to right, they are:
Surveyor 3 (USA, 1967) one of a series of probes that carried out chemical tests on lunar soil;
Luna 13 (USSR, 1967) which tested the hardness of the Moon surface;
Luna 16 (USSR, 1970) a larger machine which not only gathered a sample of moonsoil, but also sent this off back to Earth; and Lunokhod 1, a Moon rover that emerged from the even larger Luna 17 (USSR, 1970) and trundled about the surface of the Moon, sending back television pictures.

What the Machines Discovered

Astronomers, with their powerful telescopes, knew a great deal about the landscape of the Moon long before the first spacecraft rose from Earth. But the Moon always turns the same face towards us. The early Moon orbiters were able to show astronomers what they had never seen before – the far side of the Moon.

The early landers measured the Moon's climate, which consists entirely of month-long, scorching hot lunar days, followed by equally long, lethally cold lunar nights. In fact, the temperatures on the Moon's hostile surface were found to range from 212°F in the day to –238°F at night.

The Moon machines also tested the lunar soil, and confirmed what had already been suspected, that it was completely dry and barren. This lack of water, together with the lack of air, meant that the Moon could not support even the simplest forms of life.

Below: Five of these Orbiter (USA) spacecraft circled the Moon between 1966 and 1967, taking photographs in each of their orbital sweeps, to build up a complete picture survey of the Moon's surface.

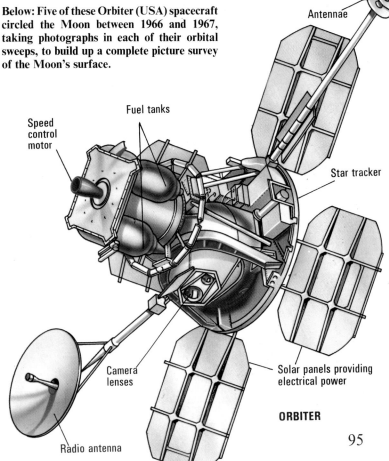

Antennae

Speed control motor

Fuel tanks

Star tracker

Camera lenses

Solar panels providing electrical power

ORBITER

Radio antenna

95

The Earth from Space

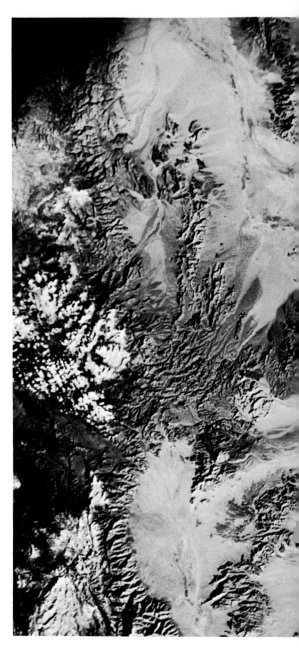

To an astronaut on the Moon, the Earth appears as a round, blue and white oasis of life against the black emptiness of space. Sadly, however, the future of life on Earth is threatened. We are using up precious resources too rapidly and we are polluting the world's rivers, oceans and air with industrial waste.

To help overcome these problems, scientists at the National Aeronautics and Space Administration (NASA) launched a special satellite in 1972. It was called Earth Resources Technology Satellite (ERTS). Three years later it was joined by a second satellite, and the two satellites were renamed Landsat 1 and Landsat 2. Landsat 1, originally intended to last one year, actually continued to work for five and a half years. Landsat 3 was launched in 1978.

Cameras Focused on the Earth

Landsat orbits the Earth in a polar orbit – that is, it circles the Earth in a north-south direction, passing over the two poles. Orbiting at over 560 miles above the surface, it completes each orbit in 103 minutes and scans a strip 115 miles wide. Meanwhile, the Earth spins below Landsat at right angles to its orbit, and the satellite scans the whole of Earth's surface once every 18 days.

Landsat carries an array of cameras and scanners that can record images in red, blue or green light and can also detect infra-red radiation. These images show up features and events on the Earth's surface that are otherwise very difficult to detect.

Above: An infra-red photograph of the mouth of the Colorado River as it flows into the Gulf of California. Although this photograph was actually taken from the Apollo spacecraft, it is the kind of view that can be seen from a satellite. Infra-red photographs can indicate the health of vegetation. Healthy foliage shows up as red, whereas the foliage of diseased or pest-infested plants appears brown. Differences between the river water and the sea can also be seen; the river water shows up as a paler blue.

Left: An astronaut's view of the Earth. The beautiful blue and white coloring of our planet is due to the presence of water. The blue oceans cover nearly three quarters of the Earth's surface. Heat from the Sun causes water to evaporate and form the white clouds. Beneath the clouds and between the oceans, continents can be distinguished. In this view of the Earth you can clearly see the outline of Africa and the Arabian peninsula.

96

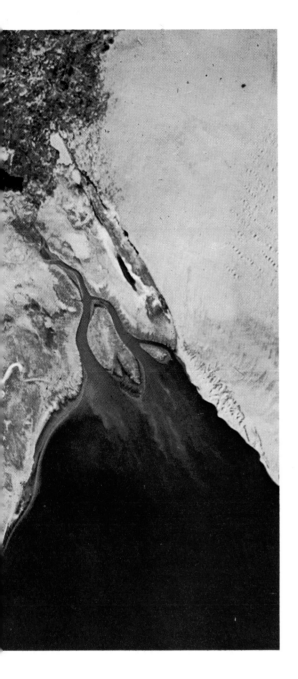

Finding Useful Materials

Landsat is making it possible for geologists to find new deposits of minerals. For example, infra-red photographs have revealed the presence of lithium-bearing and potassium-bearing rocks in the High Andes of Bolivia. The existence of minerals can also be revealed in other ways. In the jungles of the Amazon Basin, where traditional methods of prospecting are expensive and extremely difficult, Landsat has been able to locate the exact positions of tin-bearing granite rocks. The soil around these rocks is preferred by particular species of plants and Landsat pictures show slight differences in color where these plants occur.

Landsat can also help locate petroleum and underground water supplies. Satellite pictures showed geologists that some of the Alaskan petroleum deposits were much larger than had previously been thought.

Checking Pollution

Landsat has also been used to monitor pollution. Photographs have shown how smoke from industrial plants feeds into clouds and causes changes in the weather over 90 miles away from the source of the pollution. Infra-red photographs can detect chemical pollution in coastal waters and Landsat can also be used to monitor the movements of large oil slicks.

Landsat and Agriculture

Satellite pictures are also proving of great benefit to the world's farmers. Landsat can provide information about the moisture content and fertility of the soil. It has shown how controlled grazing of hot, dry areas can lead to reclaiming desert for agricultural use. It can also distinguish between different plant species and indicate whether they are healthy or not. Such information helps scientists improve methods of controlling pests and diseases.

Right: A photograph taken by Landsat 1 (ERTS) in 1973. It was produced by scanning the area in different colors, using instruments called multispectral scanners. The information was then processed electronically to produce this multicolored picture. Such pictures can provide scientists with a great deal of information about the Earth's surface. For example, they reveal at a glance the complicated arrangements of mountain chains and can also show hitherto unknown geological features.

97

News by Satellite

Above: One of the problems of keeping up communications with a satellite is pinpointing and keeping track of such a tiny object as it moves far out in space. To achieve this pinpointing and tracking, powerful and accurate radio antennas, very like radio telescopes, are often used. This array of eight giant antennas is in the Crimea region of the Soviet Union.

Below: Another way of tracking artificial satellites far away in space is the use of special cameras, which photograph a satellite as a streak of light in the night sky. The big camera in the picture is at Woomera in Australia, where the nights are often very clear and so are suitable for satellite photographing.

When you watch a television program today, you may see pictures coming "live" from places many thousands of miles away. In a news program, for example, two interviewers, one in America and the other in Britain, may be talking face to face about some important item of news.

We now take this kind of thing for granted, yet only 25 years ago no owner of a television set could have expected to receive pictures from as far away as the other side of the Atlantic.

Then, in 1962, the communications – or news – satellite called *Telstar* was launched into orbit from the United States. Transatlantic television pictures now became possible because communications engineers could now bounce television or radio signals off this useful satellite from one country to the other. In 1965 the Soviet Union launched a rather similar news satellite, called *Molniya*, for communication over the vast land distances of the Soviet Union.

Poised in Space

News satellites such as Telstar had one fairly big disadvantage. Engineers could not use such a satellite at just *any* time for the sending of a long-distance message or program. They had to wait for the news satellite to come into the right position in the sky, as it orbited the Earth far above their heads. When it passed on, over "the horizon" to the other side of our world, it could no longer be used for communications.

This problem was solved by the United States in January 1967 with the successful launching into orbit of an improved news satellite called *Intelsat*. This satellite was positioned in space to orbit Earth around the equator in exactly 23 hours, 56 minutes. As this is just the length of one Earth day, Intelsat remained directly overhead for the use of the engineers.

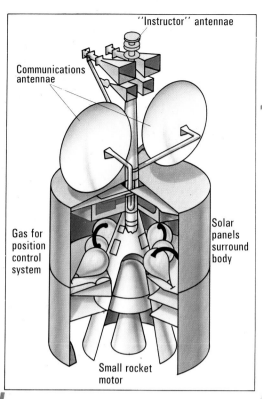

"Instructor" antennae

Communications antennae

Gas for position control system

Solar panels surround body

Small rocket motor

Above: An Intelsat news satellite, cut away partly to show its internal details. Its small rocket motor is fired after launching from Earth to place the satellite in exactly the right position in space. Electrical power for any further movements is provided by solar panels, which make electricity from sunlight.

Above: In Britain, signals or programs sent out from other countries and relayed or passed on by Intelsat are often received by this giant antenna. It is one of three operated by the Post Office at Goonhilly Downs in Cornwall, England.

Atlantic Ocean

Indian Ocean

Pacific Ocean

INTELSAT
The most modern news satellite system is that provided by such satellites as Intelsat. There are now three Intelsat satellites poised in space, one over each of the three biggest of Earth's oceans, so that their signal beams together cover the whole world surface.

Left: The control room at Goonhilly Downs receiving station shows some of the special electronic equipment needed for satellite communications.

Weather Satellites

Weather forecasting has become an essential part of modern life. To many people, such as sailors, aircraft pilots and farmers, accurate knowledge of future weather is vital. And weather conditions can change dramatically in just a few hours.

Weather depends on many things. Among these are the wetness or dryness of the air and its temperature, the presence of fronts (areas where warm air meets cold air), and the movements of depressions (areas of low air pressure) and anticyclones (areas of high air pressure).

A weather forecaster must collect and weigh up all this information in order to be able to say what kind of weather is coming to any one area. If he could stand above the clouds and observe the world's weather systems from there, then of course he could tell much more certainly and accurately what weather was in store for a particular area. Weather forecasters of today very nearly do just this – except that it is a robot or weather satellite above the clouds that does the measuring for them. It photographs and marks the progress of weather systems across the Earth's surface and transmits the information back to controllers.

The first weather satellites were put up into orbit around the Earth as long ago as 1960. The *Tiros* satellites, as these first weather-robots were called, sent back to Earth detailed pictures of clouds, that helped with more accurate weather forecasting. Later in the 1960s the *Nimbus* weather satellites, also from the United States, kept up a 24-hour-a-day watch on the weather from space. Modern weather satellites are equipped with special cameras that enable them to see weather details on both the dark and light sides of Earth.

Right: Artificial satellites, orbiting Earth at a height of about 435 to 620 miles, can take photographs which show many useful details of the Earth's surface and atmosphere. This satellite photograph shows an area of western Europe, from the Mediterranean coast in the south to Britain in the north. Among the white areas are the snow-capped peaks of the Alps and stretches of cloud.

Below: A weather satellite passes high over a giant cloud formation in the shape of a spiral, or cyclone. Weather satellites have proved very useful for recognizing violent weather, such as typhoons, well in advance so that steps can be taken to protect life and property. These satellites are often made to orbit Earth in a north-south direction. Because the Earth rotates in a west-east direction, this means that the weather satellite passes over different strips of the Earth's surface on each of its orbital revolutions, so eventually it covers the whole globe, before beginning all over again.

War by Satellite

Right: An artist's idea of aggression in space, perhaps in AD 2,000. A military satellite of one major country, armed with a powerful laser weapon, uses this to destroy a satellite of a rival power. Most likely, these two fighting satellites are unmanned, so that no astronauts are killed or injured in the conflict. The operators of the two satellites are hundreds or thousands of miles away, in their control rooms back on Earth, from which they play this latest and most dangerous of war games. For if a powerful laser beam can be used to destroy machines in space, then even more powerful weapons in space could perhaps be turned on to Earth, to destroy vast numbers of human lives.

For years there has been great rivalry between the USA and the USSR to put up more and better spacecraft. This is often called the space race. The *arms race* is the name of a more sinister competition going on, also mainly between these two powerful countries. It means the building of more and more modern weapons, each more destructive and deadly than the last, with each of the countries trying to scare and outdo the other.

For a short time, the space race did involve friendly co-operation between the USA and the USSR. The best known example of this is the 1975 meeting in space between the Soviet Soyuz and American Apollo spacecraft, when the American and Soviet astronauts linked up their spacecraft in orbit, shook hands, and climbed into one another's cabins.

Unfortunately, there have been no further demonstrations of this kind of friendliness and co-operation since 1975, although scientists all over the world, including some Americans and some Russians, do still manage to exchange useful information. But on matters concerned with the space race, the "space giants" have stopped talking – and one of the reasons is that the space race is becoming part of the arms race.

Spy Satellites

For several years now, both nations have been putting up satellites whose business it is to spy out details of the other country's attacking systems and defenses. For example, these satellites look for evidence of the siting and the moving around of giant ICBM missiles with their terribly destructive nuclear warheads. The satellites can do this because they are fitted with the very latest in long-range cameras, which can see and magnify details hundreds of miles away, even in the dark.

So far you might think that this is not too dangerous for mankind, because these spy satellites are at least not weapons of war in themselves. However, they could be made so, quite easily. Both the rival world powers may soon be capable of putting up large and powerful weapons into space, to turn a harmless spy satellite into a long-range, remote-controlled weapon.

Weapons in Space

In the USA, such a dangerous advance in the arms race has been made possible by the early success of the space shuttle. The shuttle is capable of carrying up heavy loads of equipment into space, then returning to Earth for more. So it is already possible for the USA to start building the latest kinds of weapons in space. These could include the sort of laser shown in the picture opposite, which would be capable of destroying an enemy satellite.

The USSR does not yet have a space shuttle, but it does have very advanced space rockets, together with a lot of practice in manning the Salyut space laboratory over long periods of time. So it, too, can already build weapons in space, by taking their parts up in space rockets, then putting them together, using the space laboratory as a weapon workshop.

Are weapons being built in space already, or have the "space giant" governments and military men not yet made up their minds? Although such a matter is vitally important to all of us, it is also one of the most closely guarded military secrets, and only a few politicians and generals are likely to know the answer.

Below: A laser beam in action. Lasers are very useful devices, for example in welding together metals and for carrying out certain very precise surgical operations. But like many modern inventions, they can also be used for less peaceful purposes. The latest lasers can be much more powerful and far-reaching than the one shown. Lasers of the near future will be so powerful that they may find uses as weapons, in the way shown by the other picture.

Exploring Mercury and Venus

In the age before space travel, no really good maps could be made of the planet Mercury. Besides being a small world, it is so close in to the Sun that even powerful telescopes on Earth show very little reliable detail. Before astronomers could be sure what its surface really looked like, they had to wait for a space probe to be sent out from Earth which was able to fly past and photograph the Sun's innermost planet.

The Mariner Probe

This was the American probe Mariner 10, which first flew past Mercury in March 1974, on its way from a flypast of Venus. This very successful planet-mapping probe then went on to fall into an orbit around the Sun. It returned twice to take more photographs of Mercury, in September 1974 and March 1975, before its instruments stopped working.

The many good photographs taken of Mercury by Mariner 10 show clearly an arid pockmarked little world very like the Moon, but even hotter on

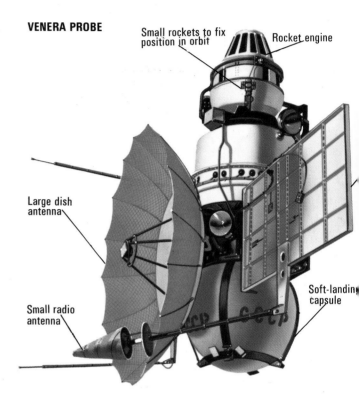

VENERA PROBE

Small rockets to fix position in orbit

Rocket engine

Large dish antenna

Small radio antenna

Soft-landing capsule

Above: The only space vehicles yet landed on the planet Venus are Venera probes from the Soviet Union. In the successful Venera probes the larger part, with dish antenna and solar panels, stayed in orbit around Venus, while the smaller capsule floated down by parachute through Venus's acid-vapor clouds to its hot, barren surface.

its sunlit side, and of course completely unsuitable for life. Mariner 10's instruments also discovered that Mercury, like Earth but unlike the Moon, has a magnetic field. Perhaps this shows that Mercury has a larger iron core than that of the Moon (see page 56).

Exploring Venus

Mariner 10 flew past and photographed Venus in February 1974, sending back clear pictures of the swirling white clouds that reflect sunlight from Venus to make it the most brilliant of starry objects in our night skies. But the same clouds completely hide from space all other surface details of Venus.

Left: The American probe Mariner 10 sent back many detailed pictures of the planet Mercury, as it flew by a total of three times in 1974 and 1975. This picture is built up from overlapping strips of photographs of the hot little world's surface.

Several years earlier, in 1965 and 1967, the Russians had landed two space probes on the surface of Venus, floating one down safely by parachute. In Venus's blisteringly hot, acid atmosphere, however, neither probe lasted long enough to send back useful information. Success came to the Russian Venus explorers eight years later, when the two probes Venera 9 and 10 made safe "soft" landings in October 1975. Although even these tougher probes sent back information only for a few hours, this was enough to give some idea of what the surface of Venus really looks like. Pictures from the Venera probes revealed a barren landscape littered with smooth rocks. Surface temperatures measured were about 840°F, or just below red heat. The pressure of the Venus atmosphere measured far more than that of Earth, and light was murky below a high cover of acid-vapor clouds that completely shut out the Sun.

In 1978 the Pioneer Venus project put two craft into orbit around Venus, and landed four surface probes. These too have provided us with valuable information about, and photographs of, this strange and hostile world.

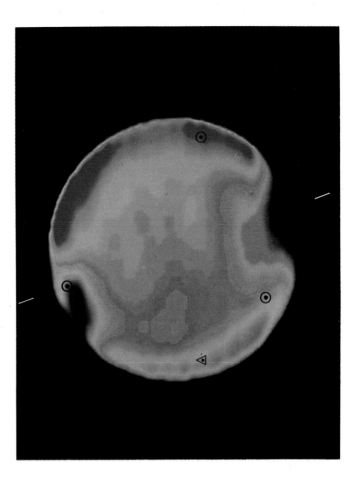

Below: The Pioneer Venus orbiter also took pictures of Venus' surface with the aid of radar waves, which can penetrate the thick cloud cover of the planet. The radar scanning provided an artist with the information on which this view of a huge rift valley on Venus is based.

Above: When the Pioneer Venus probe entered the thick atmosphere of Venus, it took pictures of the planet with the aid of long-wave, infra-red radiation. These were used to make up this map.

Exploring Mars

Mars is the most explored of Earth's fellow planets. But it has not yet been explored as thoroughly as the Moon, because no human has yet set foot on Mars. Perhaps a manned expedition may be sent out to Mars before the end of this century, and astronauts may be televised walking about on the surface of the red planet, as they were on the lunar surface in the early 1970s. Possibly, though, modern advances in robot technology will make this unlikely. The first explorers from Earth to range far and wide on the Martian surface may be very sophisticated robots, far less likely than human beings to come to harm in Mars's hostile climate.

So far, in the first 20 years of the space age, about a dozen unmanned spacecraft from Earth have sent back information about Mars. Several have done this as they passed close by the planet, before carrying on into space, often to be lost in a great orbit around the Sun.

Some other spacecraft have been intended for orbit around Mars itself. Of these, the American probe, Mariner 9, was the most successful. Launched in November 1971, this took more than 7,000 photographs while in Mars orbit. Altogether, these photographs provide us with a surface map of Mars as detailed as many we have of parts of the Earth's surface.

What Mariner Saw

Some details of the Martian surface proved to be very impressive indeed. Although Earth has three and a half times the surface area of Mars, the highest mountains of Mars are three times as lofty as those of Earth. Olympus Mons, largest of all, is a giant volcano 15 miles high and 373 miles across.

A second huge feature of the Martian surface is Hellas, a circular patch thought to be a mountainous area until Mariner spacecraft showed it to be a low-lying plain, and for some reason the least pockmarked region of Mars. Elsewhere craters were seen to be plentiful, except in the polar regions where the surface of Mars is covered with water ice and carbon-dioxide ice. A third impressive feature of the Martian surface was the great rift valleys, the largest of which, lying between Hellas and Olympus Mons, is 2,485 miles long.

This photograph of the crescent Mars was taken by the U.S. Viking Orbiter 2 in 1976. It clearly shows a plume of ice blowing from the peak of the volcano called Ascreaus Mons, at the top left of the picture. The Viking Orbiter is shown on the opposite page.

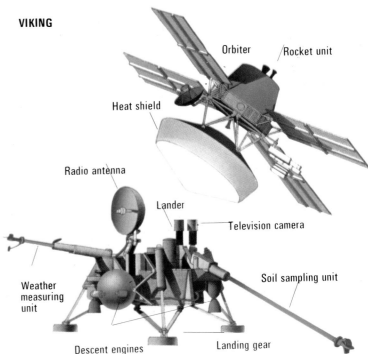

Orbiter

Rocket unit

Heat shield

Radio antenna

Lander

Television camera

Weather measuring unit

Soil sampling unit

Descent engines

Landing gear

Above: The Mars landers that sent back the surface pictures on this page were those from the spacecraft Viking 1 and 2, which landed on the red planet late in 1976. The picture shows both parts of the Viking craft, the one that stayed in Mars orbit and the other that descended to the Martian surface.

Above: A Viking Mars lander sits alone in a barren, sandy red landscape. This is not Mars itself – because no one would be there to take the photograph! The photograph was in fact taken during tests of Viking in a desert of the USA.

Right: This photograph is of the actual Martian surface. Taken by a Viking lander, it shows how barren and lifeless the Martian surface really is. The Martian sky looks orange because of the red dust floating in the thin Mars atmosphere.

The Mars Landers

In July and September 1976 the American spacecraft Viking 1 and Viking 2 touched down safely on the surface of Mars, took photographs and measurements, and tested the soil of Mars for signs of life. The pictures sent back by these Mars landers were fascinating but not altogether surprising. They showed the red, stony, desert landscape so often imagined by science fiction. In the Martian daytime, the sky over this red landscape was pink, probably with dust.

Tests for living organisms were at first puzzling, but in the end perhaps disappointing. Oxygen found in the Martian soil might have proved a sign of life, if only of a primitive kind. But it seemed more likely to be the result of a purely chemical reaction in the soil.

The Pioneer Program

In March 1972, only a few months after Mariner 9 had been sent off to orbit Mars, the first space probe aimed towards the outer, giant planets was launched, also from the United States. It took Pioneer 10, as this probe is called, until December 1973 to approach the first of its targets, Jupiter. It had taken 21 months to travel across 62,137,100 miles of space to arrive at the biggest of the Sun's planets.

On its way Pioneer 10 had to pass through the wide belt of asteroids, or minor planets, that sprinkle space between the orbits of Mars and Jupiter. There were some fears that a rocky or pebble-like asteroid might destroy the spacecraft or put it out of action. But Pioneer 10 survived this passage with all its instruments still working, and so did its successsor Pioneer 11, which was launched just over one year later.

Another hazard loomed up as the Pioneer spacecraft approached near Jupiter. By "near" is meant about half the distance that the Moon is from Earth. At this distance the spacecraft were bathed in intensely strong belts of radiation trapped in the planet's enormous magnetic field. Scientists on Earth watched their passage fearfully, but again both spacecraft survived.

The sheer strength of this radiation was one of the major discoveries of the Pioneer craft. It showed that Jupiter gave off much more energy than it received from the Sun, and so, deep down in its inside, must be an intensely hot and active planet. The Pioneer spacecraft, while passing quickly across Jupiter's vast face, also measured the planet's strong and peculiar magnetism, and sent back pictures of its striped, gassy surface in fascinating detail.

From far-away Earth, astronomers had often guessed that the Great Red Spot on Jupiter's surface might be some kind of an island, liquid or solid, that sank into the gassy surface every now and then when it became less clear or distinct. But the Pioneer spacecraft, from their much more close-up view, were able to show that the Great Red Spot is really a gigantic whirling storm, larger in surface than the whole of the Earth, and that probably its occasional brightening or fading is caused by the storm stirring up or dying down.

Pioneer 10 and 11 did not stop at Jupiter, nor were they pulled into orbit around the great planet. Instead, they were aimed cunningly to use Jupiter's powerful gravity to push, or sling, them still farther out into the Solar System.

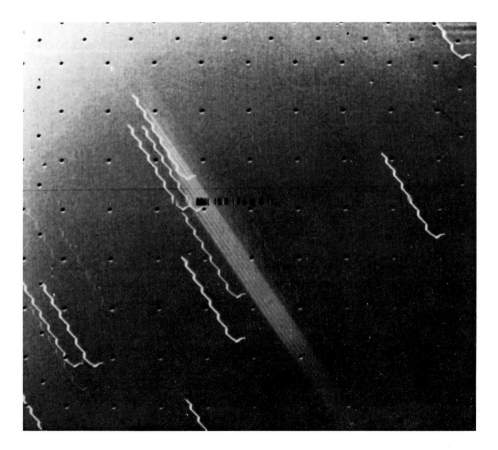

In March 1979, a Voyager spacecraft was approaching Jupiter's vast surface when its camera briefly detected a streak. This was later identified as part of a ring around the giant planet. Much fainter than Saturn's rings, Jupiter's ring is apparently made up of rocky boulders. Although the Jupiter ring is very wide, its boulders are thinly spread, so that the ring only becomes clearly visible, even at near distances, when viewed edge-on.

Right: This diagram shows the space flight of Pioneer 11, the first space probe to reach Saturn. First, Pioneer 11 was sent to Jupiter. Nearly two years later, it encountered this giant planet. Pioneer 11 passed close by Jupiter and was slung farther onwards into space by Jupiter's powerful gravity. In five years more, Pioneer 11 reached Saturn, and was again catapulted onwards into space. The marks on each planet's orbit show its position at the date given.

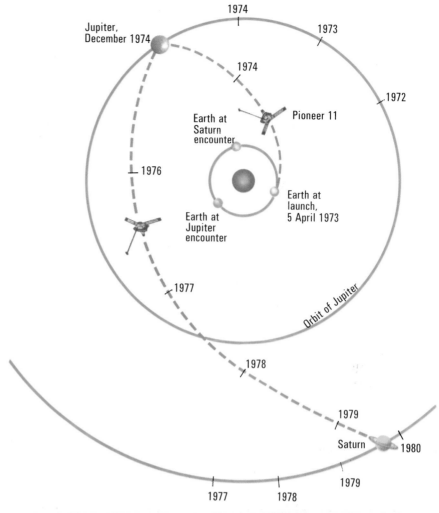

1974
1973
Jupiter, December 1974
1974
1972
Earth at Saturn encounter
Pioneer 11
1976
Earth at launch, 5 April 1973
Earth at Jupiter encounter
1977
Orbit of Jupiter
1978
1979
Saturn
1980
1977   1978   1979

This picture of Saturn's rings and its moon Tethys was taken by one of the Pioneer space probes in August 1979, from 585,922 miles away.

Voyager to the Outer Planets

The slingshot effect of Jupiter's pull of gravity sent the spacecraft Pioneer 10 hurtling on towards the outer limits of the Solar System. Pioneer 11, following a year later, had a more immediate destination – to fly past the next giant planet, Saturn. But it was not to achieve this for another five years, in 1980.

Meanwhile, in 1977, two still more sophisticated spacecraft were launched from the Cape Canaveral rocket base in the southern United States. Voyager 1 and 2, as they are called, were also aimed at the giant planets, and with their improved instruments they were expected to provide even more useful information about them than that of the Pioneers.

One after the other, in spring and summer 1979, the Voyagers flew past Jupiter. They sent back many detailed pictures not only of Jupiter itself but also closeup shots of several of its larger, inner moons. Then, slung onwards by Jupiter's gravity even faster than the earlier Pioneers, the Voyagers reached Saturn and its rings a mere one and a half years later, in November 1980 and August 1981. After flying by Saturn – passing right through its rings – Voyager 1 was then catapulted onwards by Saturn's gravity into farther, uninhabited regions of space. Voyager 2, though, headed for a rendezvous with the next farthest-out planet Uranus. It reached Uranus in 1986 and will pass by remote Neptune in 1989.

The discoveries made about Saturn by Pioneer 11 and the Voyagers are many and exciting. They include several new moons and many new rings, one ring being coiled like the strands of a rope.

This painting shows the flight path of the Voyager spacecraft through the Solar System. Launched from Earth – the third planet out from the Sun – its carefully planned trajectory took it past Jupiter before it journeyed on to photograph Saturn, the ringed planet. The photographs which Voyager sent back to Earth, two of which are shown opposite, have added enormously to our knowledge of these giant planets.

A Voyager shot of Saturn, taken from a distance of 46,602,825 miles. Also visible are five of Saturn's 24 or so moons. They are Titan (the largest), Dione, Tethys, Mimas and Enceladus.

Like Pioneer 11, but in still more detail, the Voyager spacecraft showed that Saturn's rings are much more numerous and complicated than had been suspected. This Voyager picture of Saturn's rings is taken from the dark side of the giant planet, at a distance of 434,960 miles.

Skylab

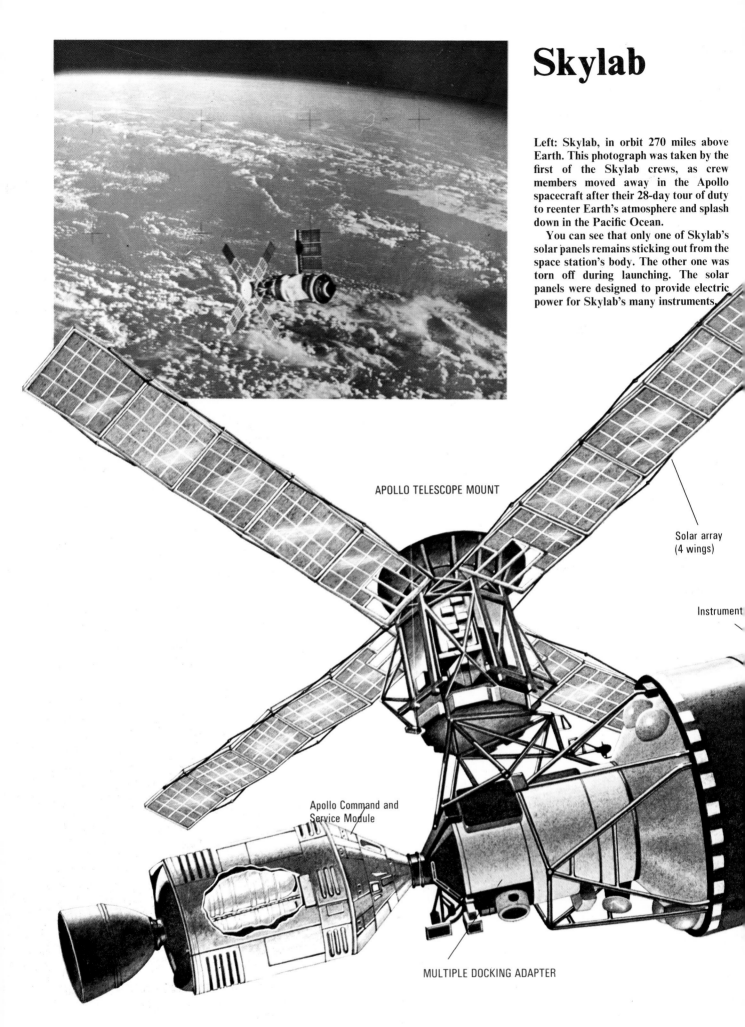

Left: Skylab, in orbit 270 miles above Earth. This photograph was taken by the first of the Skylab crews, as crew members moved away in the Apollo spacecraft after their 28-day tour of duty to reenter Earth's atmosphere and splash down in the Pacific Ocean.

You can see that only one of Skylab's solar panels remains sticking out from the space station's body. The other one was torn off during launching. The solar panels were designed to provide electric power for Skylab's many instruments.

APOLLO TELESCOPE MOUNT

Solar array (4 wings)

Instrument

Apollo Command and Service Module

MULTIPLE DOCKING ADAPTER

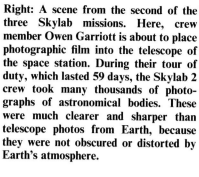

Right: A scene from the second of the three Skylab missions. Here, crew member Owen Garriott is about to place photographic film into the telescope of the space station. During their tour of duty, which lasted 59 days, the Skylab 2 crew took many thousands of photographs of astronomical bodies. These were much clearer and sharper than telescope photos from Earth, because they were not obscured or distorted by Earth's atmosphere.

Below: A closeup of Skylab, showing the crew quarters, and below left, the Skylab badge.

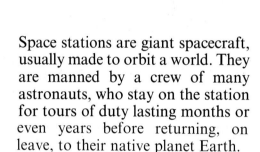

Solar array wing ripped off at launch

Crew quarters

Refrigeration system radiator

ORBITAL WORKSHOP

Experiment compartment

Solar array wing

Space stations are giant spacecraft, usually made to orbit a world. They are manned by a crew of many astronauts, who stay on the station for tours of duty lasting months or even years before returning, on leave, to their native planet Earth.

This, of course, is a picture of the future – but only of the near future. Already, during the past 10 years, Salyut (USSR) and Skylab (USA) space stations have been orbiting hundreds of miles above Earth. Their three-astronaut crews have stayed up in them for as long as several months at a time, carrying out scientific experiments and taking a detailed look at the Universe in the clear, empty space surrounding their orbiting home.

Skylab, shown on these pages, consisted mainly of a large tube which was part of a Saturn V rocket. This empty rocket tube, or stage, was converted into living and working quarters, which were occupied by the three Skylab crews, the last of which stayed up in space for 84 days.

Skylab itself drifted out of orbit and burned up in the Earth's atmosphere in 1979, scattering flaming debris over Australia.

113

Daily Life in Space

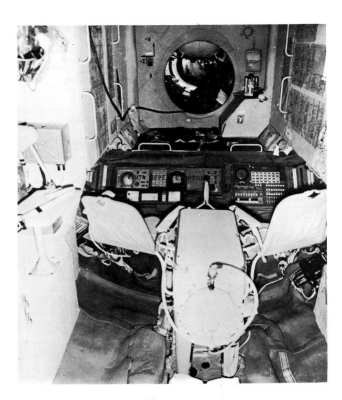

The first astronauts stayed up in their strange, weightless world for at most a few days. On later moontrips this period was extended to weeks, and in the orbiting Skylab and Salyut space stations astronauts have stayed in space for a matter of months. How did these longer stays affect the astronauts?

In the days before men had left for space, there was some fear that meteorites and harmful radiation would make space too dangerous for life. But the Skylab and Salyut crews survived months-long exposure to these hazards, without injury. Weightlessness in orbit was a more mysterious problem. All life on Earth, including our own, is "designed for gravity." So would not prolonged weightlessness

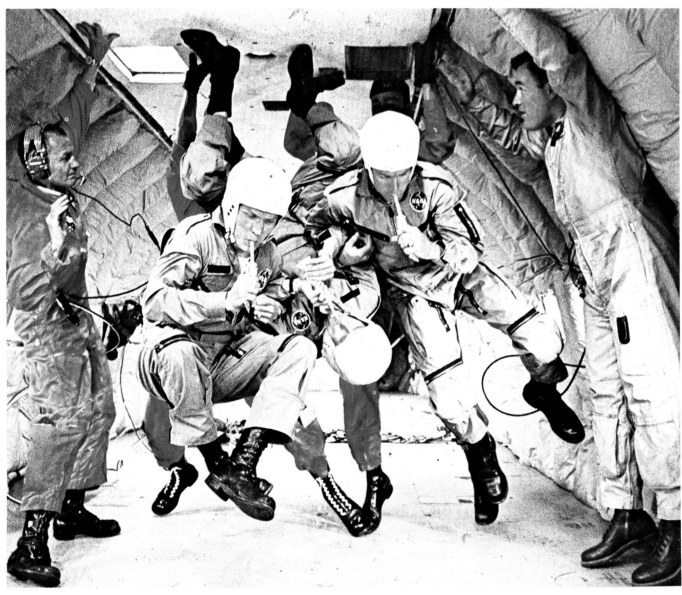

Above: Charles Conrad takes a bath in Skylab. He needs this complicated-looking arrangement to do so. Water would soon escape from any ordinary type of bath under the weightless conditions and would fill the whole spacecraft with droplets.

Right: Topsy-turvy in orbit. The two astronauts shown are checking instruments in the Skylab space station. They are Joseph Kerwin (top) and Charles Conrad (bottom).

Opposite above: A view inside the control room of the first Salyut space station, put into space by the USSR in 1971. Beyond the astronaut's seats and the control panel is an open hatchway leading to the spacecraft that brought up the astronauts and docked with, or attached to, the space station.

Opposite below: Astronauts undergo training in weightless conditions (in an airplane diving from a high altitude). Here they are practicing drinking under zero-gravity.

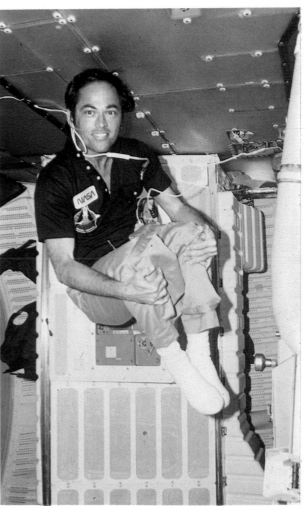

Right: Weightless in orbit. Robert Crippen, pilot of the first space shuttle, floats in his spacecraft Columbia, which he later helped to bring back to Earth base.

cause trouble, for example with the body's flow of blood – which certainly is influenced by the pull of gravity?

Doctors found that astronauts who had served long periods of space duty did show some body changes, including changes in blood flow. But in most cases, the astronauts soon returned to normal after a short while back on Earth. In other words, their bodies had adapted, or changed, first to suit the weightless space conditions, then back again to suit the weighty Earth conditions. This was true also of other creatures than ourselves. For example, spiders taken up in the spacecraft spun their webs and lived normal enough lives, without the pull of gravity.

The Space Shuttle

The conquest of space is people's greatest feat of technology. It has also been their most expensive and wasteful project. In the United States, every time a giant, 364-foot Saturn rocket blasted off, tens of millions of dollars went with it, because nothing except the little capsule containing its crew was ever recovered.

But a new, less wasteful pathway into space has been designed by American rocket engineers. This is the space shuttle, which marks the second great stage of the space age. Briefly, it is a system for sending up a large cargo carrier into orbit, then bringing it back safely for use in further space flights – perhaps as many as a hundred altogether.

The first two space flights took place in 1981. Later, more missions were run, including both two-man and four-man crews, and in 1983, the first U.S. woman in space. In January 1986 the shuttle *Challenger* exploded after takeoff, killing the seven crew members. No further shuttle flights took place until September 1988.

How the Space Shuttle Works

The space shuttle blasts off with the power of the orbiter's rocket engine, plus that of two slimmer booster rockets. The boosters are solid-fueled, and after they have done their boosting, fall away and are recovered by parachute.

By this time, the orbiter is 25 miles high in the atmosphere, still sitting on its large, rocket-shaped fuel tank. Using the liquid fuel in this tank, the orbiter blasts on upwards into space. At a height of more than 62 miles the large fuel tank, now empty, falls away to burn up in the Earth's atmosphere – the only wasted part of the space shuttle.

Coasting on (without using its engine) the orbiter makes its first trip around Earth, usually the first of many orbits. While in orbit, its crew members carry out any special tasks assigned to them. When the time comes for the orbiter to return to Earth base, they fire the orbiter's rockets once more, this time to slow the orbiter down and so to make it fall towards Earth. After its hot passage through the atmosphere, the orbiter behaves like a giant glider, and is brought down to base by its pilot. It may have carried a cargo of many tons into space, but now it is empty – and ready for the next payload.

Above: The space shuttle blasts off with all the power of its main rocket engine and two booster rockets. High up in the atmosphere, the boosters will fall away. Even higher, in space, the large fuel tank will be detached, leaving the orbiter free to carry on into space and eventually to glide back to Earth base.

Left: High above Earth, the orbiter releases a large satellite into space. As it does so, it reveals the impressive roominess of its cargo bay. The orbiter is as big as a medium-sized airliner and can carry bulkier loads up into space than were possible even with the giant Saturn rockets. And unlike them, the orbiter returns to Earth for more.

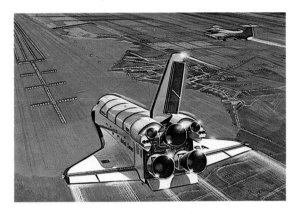

Left: Back to Earth from space. On its short, delta wings, the orbiter makes a fast glide-down along the runway of its return base. In this rear view you can see the three big nozzles of its main rocket engine, and the two smaller nozzles of the rockets used to maneuver, or move about, the spacecraft in orbit. On its return trip to Earth, all these rockets are shut off.

Below: Columbia, first of the space shuttle orbiters, touches down under the guidance of its pilot. The event is an historic one for space flight, because it marked the end of the first, successful return trip for the new space ferry. In years to come, Columbia and its sister orbiters will carry heavy loads up into space and return for more. This will lead to such things as the construction of giant space stations and to space tourism.

Building a Giant Space Station

With the aid of the space shuttle, giant space stations may soon be built to orbit the Earth. This picture shows a scene of the near future, during the construction of such a station. It is being built at a height of some hundreds of miles above the Earth's surface.

One large part or "city block" of the space station has been completed already. This block includes a terminal, or parking place, for the space shuttle's winged orbiter, which has recently arrived from Earth. In its roomy cargo bay, it has brought up a many-ton load of long metal sections and other equipment for the space station, which is being put together in space rather like a giant Erector set.

In some ways, putting up a building in space is easier than putting one up on the Earth's surface. In orbit, there is no force of gravity to give any object weight. For this reason, not even the largest of metal sections, or other part that would weigh heavily on Earth, takes anything like the same amount of energy to put into position – once it has been brought up by the shuttle.

You can see that a large part of the space station's skeleton has already been put together from the metal sections. Another kind of metal unit can be provided by the empty tubes of large rockets which perhaps could act as living and storage spaces during the earlier stages of construction. of construction.

When the space station has been completely covered in heat-proof and cold-proof layers, then the great building in space will begin to turn into a place fit for many human beings to live in.

Many reasons exist for putting up a large, manned space station. The first, perhaps, is to create a place where scientists can work, together with all the special equipment that they need to carry out useful research. Scientific research stations would quickly advance such branches of science as astronomy – and the science of space travel itself. On the next page we will talk about several other good reasons for giant space stations.

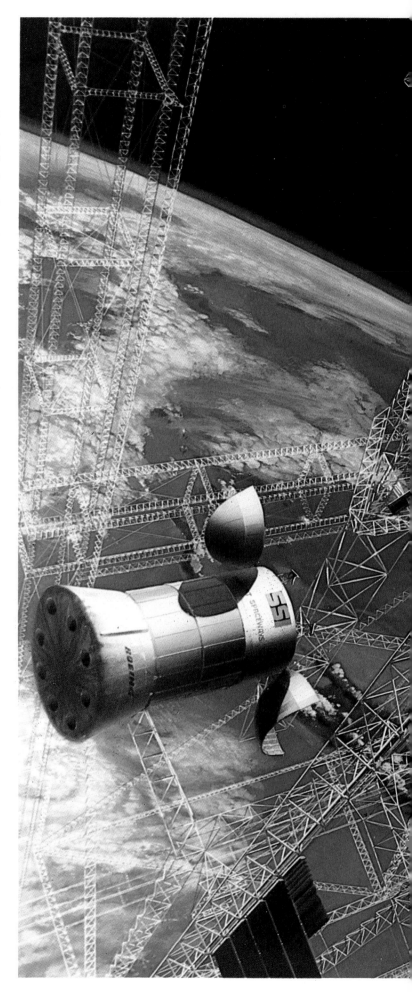

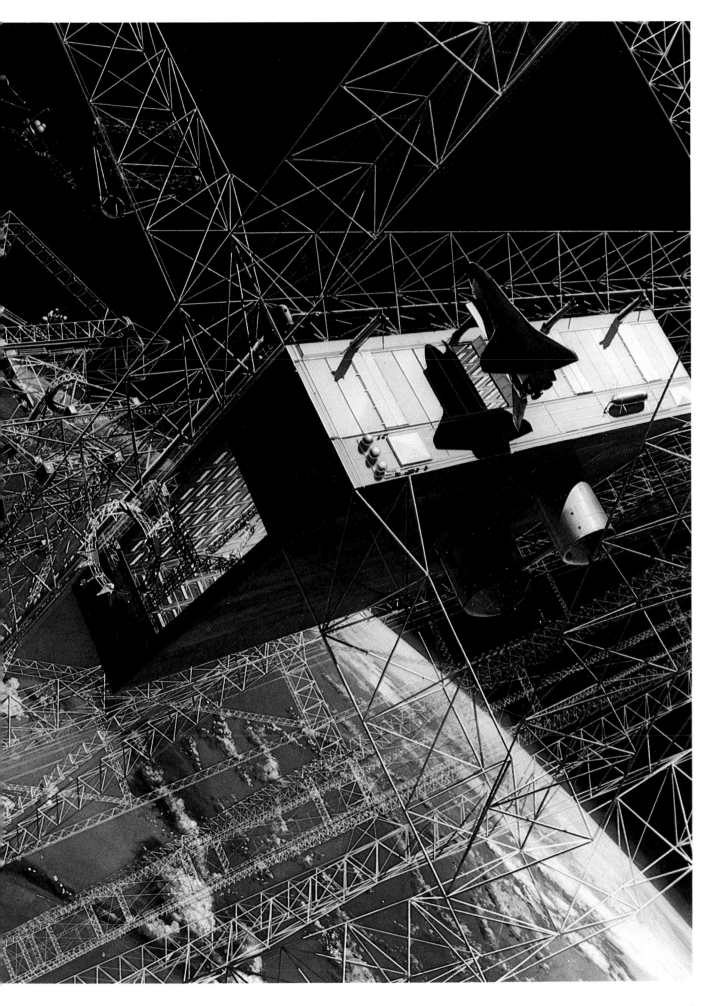

Cities in Orbit

Whole cities of human beings living permanently up in space may seem a fanciful notion. Yet there are many convincing reasons why it could be a very good idea. Some of these reasons are of the doomladen or "doomwatch" kinds. With more than four billion (four thousand million) people, Earth is already becoming an overcrowded planet. Human populations are still growing rapidly and look like they will continue to do so for at least another century unless some catastrophic event, such as war or disease, reduces them.

The four billion people who live on Earth at present use up far more of our planet's limited resources than ever before. They gobble up more food – even though many millions live at starvation level. They burn much, much more of the Earth's fossil fuels – coal and oil – than ever their grandfathers did. This quickly uses up the Earth's store of fossil fuels and also creates another worldwide problem by helping to pollute the Earth's surface with poisonous industrial wastes.

Obviously, something will have to be done to save our overcrowded Earth. Space cities and space stations are one good answer because they siphon away, or remove, both populations and industry from Earth, so leaving our planet with only "Earth-sized problems" that it can deal with.

Building a Space City

A space city would be built to hold a human population of a million or even many millions. A favorite idea of how such a city would look is given by the picture. This space city is in the form of a gigantic, hollow wheel, many miles across and over a mile thick. Putting together such a huge construction in space would certainly take many years, but would be far from impossible, for two main reasons.

The first reason is that heavy weights are no great problem out in Earth orbit or in deeper space where the pull of gravity is not felt. So, moving great building sections of the city about would be less of a problem to the builders than if they were back on Earth.

The second reason is that all the effort that will be required will be made not by human builders, but by robot builders. Probably no human being will be present at all when the first space city is under construction, until people go to live there permanently. Until that time, instructions for building the great city will either have been beamed out from Earth headquarters, or will be held in the memories of the robot builders themselves. The raw materials for space building might be mined on the Moon and flung out into space for the robot builders to collect.

Living in a Space City

When the time comes for the space city to be colonized by its million or more inhabitants, these will be shipped up to it from Earth, in rocket liners similar to the one shown in the picture. Other rocket freighters will already have carried up to the space-city all the other things, and forms of life, necessary to keep human beings well fed and happy.

Trees will have been planted to provide both pleasant shade and fruit to eat. Market gardens will have been established to provide fruits, vegetables and flowers. The energy for growing all

Two views of a space city of the future, built in a wheel shape or *torus*. The central "hub" houses solar power and docking facilities, and mirrors to reflect sunlight into the living areas in the rim.

these green things will come from the same source as it does on Earth – the Sun. Not only human beings will take advantage of all this green food, but also farm animals, domestic animals, garden insects and other necessary forms of small life – and perhaps, even some wild creatures. Finally, factories and workshops will already have been built to provide the new population with useful employment, although most of the mechanical work of the city will be done by useful robots, if so desired.

Astronauts of today are obliged to live a weightless life in space, but this condition would be very inconvenient in a space city. It will be avoided by having the city spin slowly in space. This presses everything in the city towards the outer rim, by an artificial force of gravity.

Left: A view of a space city, shaped like a giant wheel turning in space. The view shows part of the wheel and one of its hollow spokes, which could contain a road, possibly together with factories and power plants. The slow spinning of the wheel would create a force that would feel like Earth's gravity to the city's inhabitants. They would get their food from farm animals and plants, as we do – all the energy for growing green plants for human and animal food would come from the Sun.

Beyond the Solar System

In the short 30-year history of space travel, human beings have reached out to the Moon. But unmanned space probes launched from Earth have already voyaged a great deal farther than this. Pioneer 10, first of the outer-planet probes, passed by the giant planet Jupiter as long ago as 1973. At the time, Pioneer 10 was traveling at the tremendous speed of 105,630 miles per hour. At such a speed, Pioneer 10 will by now have swept far beyond the orbit of Pluto, at the limits of our Solar System.

Pioneer 10 successfully completed its main task of photographing Jupiter in closeup, then carried on, to be lost in outer space. But why should contact be lost with such a far-voyaging space probe – that is, if its instruments continue to work properly, and if it does not encounter an accident such as a collision with a meteorite?

Keeping and Losing Touch

The answer to this question is all about *signaling*. At the time that Pioneer 10 was photographing Jupiter, its cameras and other working parts were being controlled by radio signals sent out from Earth. Because the probe was so far away, these controlling signals took three-quarters of an hour, traveling through space, to reach the probe. Pictures sent back by Pioneer 10 also took three-quarters of an hour to reach the television screens of the Earth controllers. So you can see that the controllers could not know whether a command had been carried out properly until an hour and a half had gone by.

Now to control a probe much farther away still, in the region of the far planet Pluto, controllers would have to wait more than 10 hours for a reply. But this is only one-half of the problem of keeping in touch. The other half is that signals grow fainter as they come from farther away. Signals coming from a very remote probe will eventually become so weak that they will no longer be detected back here on Earth, and all contact with the probe will be lost.

But this is not quite the end of the story, because, of course, space probes can be made to send back more powerful signals so that these reach us from farther away. These larger, more powerful probes would be fitted with a computer brain that would itself control the probe's instruments and flight path, without the need for time-consuming signals from Earth. Most likely, our first knowledge of other worlds outside our Sun's own system will come back to us from these intelligent probes of the future.

This photograph shows Saturn, the ringed planet, and its largest moon Titan. It was sent back to Earth from Pioneer 11, the second of the outer planet probes, which by now has traveled far beyond the limits of the Solar System.

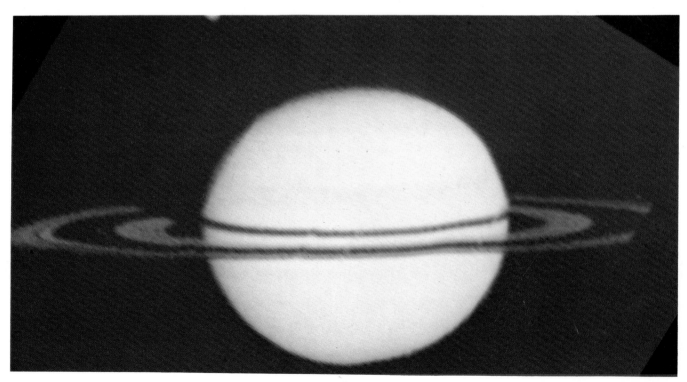

A Pioneer space probe reaches the giant planet Jupiter and its moons. Like an invisible catapult, Jupiter's powerful gravity will swing the probe onwards into outer space. With the first of the two Pioneer Jupiter probes, this happened nearly a decade ago so that by now Pioneer 10 will have journeyed well beyond the limits of the Solar System, the most far-flung of all earthly objects.

Searching for New Worlds

Of all the Sun's family of planets and their moons, only planet Earth is capable of supporting life as we know it. This is one of the messages that space probes have been sending back to us, as they carry out their missions of exploring the Solar System. Of Earth's two neighboring planets, Venus is too hot and choking, and Mars too cold and waterless a world, to favor life. The latest space probes, flying out beyond the giant planets towards the far limits of the Solar System, will encounter worlds that, whatever their interest to us, will be far too frozen ever to offer another home for mankind. In order to find new planets suitable for human life, space journeys will have to be made well beyond the limits of the Solar System, at least as far as the stars that are our Sun's nearest neighbors in space.

Journey to a Star

It is just possible that one of these nearby stars might be orbited by a planet suitable for our own kind of life. But even these nearby stars are an

To find another planet home suitable for mankind, spaceships will need to journey beyond the limits of our Sun's own system to the regions of other stars. Here, a spaceship passes close by one of these foreign suns. It has sent out a smaller spacecraft, perhaps to investigate one of the sun's planets, to see whether this is suitable for a human settlement.

If there are other intelligent beings living in our own part of the Universe, they may first get to know of us by this card, or plaque. It is carried in Pioneer 10, first of the American space probes to journey beyond the Solar System. An adult man and woman are shown against the basic plan of the spacecraft, which indicates their size. The man's hand is raised in what is hoped will be recognized as a gesture of friendship. At the bottom of the card is shown our Sun and its planets, and the path of Pioneer 10 from Earth. The symbol at the top represents the molecular structure of hydrogen – the most common element in the Universe. The starlike pattern represents the 14 pulsars of the Milky Way galaxy. Using these the inhabitants of a distant planet could identify our Sun and discover where Pioneer 10 was launched.

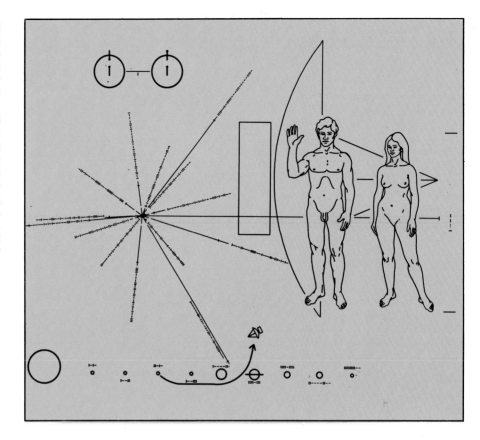

immense distance away, by earthly standards – so how long would it take a spaceship from Earth to reach them?

This question can be answered easily enough, by doing a simple sum. First of all, we know how far away the nearest other stars are – about four and a half light years. We know about how fast the spaceship seeking the new world would take off – about 24,855 miles per hour, this being the speed with which an object will escape Earth orbit for outer space.

Now that we have the starting speed and the full distance, we must decide on the rate of increase of speed, or acceleration. For the comfort of the people on the spaceship, let this acceleration be a gentle 10 inches per second per second. (For each second, the speed gets 10 inches faster.)

To reach a nearby star, the spaceship will need to accelerate until it is halfway there, then slow down, or decelerate, for the other half of its four and a half light-year journey. Then it will arrive at the star at about the same speed that it started out with. At its halfway mark of two and three-quarter light years, the spaceship will have reached a speed of about one-third that of light. Its half-journey will have taken it just over 13 years.

If the spaceship then starts to slow down at the same rate (10 inches per second per second), it will take just the same time to do the second half of its journey. Finally then, it will reach the nearby star,

and with lots of good luck, its suitable planet, in a total time of 26 years.

An even simpler sum will tell you that by the time the spaceship reaches its destination, any baby born on the spaceship, soon after it left Earth, will have grown up to be a young adult, fit and able to face the task of colonizing the new planet.

Goodbye Forever?

This is only the most favorable story of people's quest for another world. Much more probably, the space journey would last even longer, because it is more than likely that the nearest stars have no planet suitable for human life. This would mean another journey, this time of another 26 years or more. Quite possibly, the journey in search of a suitable new world would last so long that many generations of space colonists would live out their lives on the spaceship, without ever seeing another world.

Such an immense journey would be quite possible because the spaceship would be a travelling world in itself. All necessary food would be grown inside it, and it would need no fuel supplies because it would use the energy of the stars to propel it ever farther through space. Eventually, its human beings would find their new world – but they would long ago have said goodbye forever to their old one.

Dictionary

A

Absolute magnitude is a measure of a star's actual brightness, as opposed to its apparent MAGNITUDE.

Adams, John Couch (1819–1892) was an English astronomer who calculated the existence of the planet Neptune from the effects that its gravity had on the observed motion of Uranus. Neptune was discovered in 1846 by J. G. GALLE close to the position Adams had calculated.

Aldebaran is a RED GIANT star 68 light-years away in the constellation TAURUS. It is about 40 times the diameter of the Sun.

Aldrin, Edwin (b.1930) was the lunar module pilot on the Apollo 11 mission which made the first manned Moon landing in July 1969. Aldrin followed Neil ARMSTRONG onto the Moon, where the two collected samples and set up scientific experiments.

Algol is a star 82 light-years away in the constellation PERSEUS, which periodically changes in brightness. The variability is caused by a second star which eclipses it every 2.87 days.

Alpha Centauri is the nearest bright star to the Sun, 4.3 light-years away; it lies in the southern hemisphere constellation CENTAURUS. Actually, it is a group of three stars, one of which, PROXIMA CENTAURI, is slightly closer than the others.

Altair is a bright white star about 16 light-years away in the constellation Aquila, the eagle.

Altazimuth mounting is a simple form of mounting for a telescope or binoculars. The mounting allows the instrument to swivel freely up and down (in altitude) and from side to side (in azimuth).

Andromeda is a major constellation of the northern hemisphere of the sky, representing a princess of Greek mythology. It is best seen during autumn evenings.

Andromeda galaxy is a spiral-shaped galaxy of about 300,000 million stars visible to the naked eye as a fuzzy patch in the constellation ANDROMEDA. It lies about 2.2 million light-years away.

Anglo-Australian Observatory is an astronomical observatory at Siding Spring, New South Wales, Australia, jointly operated by the UK and Australian governments. Its main telescope is the 13-foot Anglo-Australian reflector, opened in 1974.

Antares is a red giant star about 400 light-years away in the constellation SCORPIUS. It is about 300 times the diameter of the Sun.

Aphelion is the farthest point from the Sun that a body such as a planet reaches in its orbit.

Apogee is the farthest point from the Earth that a body such as an artificial satellite reaches in its orbit.

Apollo project was the American man-on-the-Moon program. A total of 12 American astronauts landed on the Moon in the Apollo project between July 1969 and December 1972. The astronauts were launched into space in the three-man Apollo command module; two of them later transferred into the spidery lunar module to make the Moon landing. Apollo capsules were also used to ferry astronauts to the SKYLAB space station, and for the link-up in July 1975 with a Soviet SOYUZ.

Apparent magnitude – see MAGNITUDE

Aquarius, the water carrier, is a constellation of the zodiac, lying in the equatorial region of the sky. The Sun passes in front of Aquarius in late February and early March.

Arcturus is a red giant star 36 light-years away in the constellation Bootes, the herdsman. It is 27 times the diameter of the Sun.

Arecibo Radio Observatory, in Puerto Rico, is the location of the world's largest radio astronomy dish, 1,000 feet in diameter. The Arecibo radio telescope is slung from towers in a natural hollow in the hills.

Ariel satellites are a series of UK scientific satellites, launched by the United States. Most famous of the series was the X-ray satellite Ariel 5, launched in 1974.

Aries, the ram, is a faint constellation of the zodiac lying in the northern hemisphere of the sky. The Sun passes in front of Aries from late April to mid-May.

Aristarchus (3rd century BC) was a Greek astronomer who measured the relative sizes and distances of the Sun and Moon. He found that the Moon is roughly one-third the size of Earth, which is nearly correct; however, he badly underestimated the size and distance of the Sun. Nonetheless, he found that the Sun was considerably bigger than the Earth, and therefore made the first known suggestion that the Earth orbits the Sun, rather than the other way about.

Aristotle (384–322 BC) was a Greek scientist who maintained that the Sun orbited the Earth, along with all other celestial objects. Aristotle held that only circles, or combinations of circular motions, were allowable to explain the orbital paths of the planets. This view was eventually overthrown by Johannes KEPLER.

Armstrong, Neil (b.1930) was the first man to step onto the Moon. He commanded the Apollo 11 mission, which made the first manned Moon landing in July 1969. His first words on the Moon were: "That's one small step for a man; one giant leap for mankind."

Asteroids are lumps of rock and metal orbiting the Sun. They are also known as minor planets. Asteroids range in size from about 620 miles down. As many as 100,000 asteroids may be visible in the largest telescopes. The first asteroid to be discovered was CERES.

Astronomer Royal is an honorary title given to a leading British astronomer. Until 1972, the title was automatically awarded to the director of the ROYAL GREENWICH OBSERVATORY, but now the two posts have been separated. The current Astronomer Royal is the Cambridge radio astronomer, Sir Martin Ryle.

Astronomical unit is the average distance between the Earth and Sun; it is often abbreviated a.u. One a.u. is 92,955,803 miles.

Atlas rocket is used for launching American satellites and space probes, either with an Agena or a Centaur upper stage. Atlas rockets were used to launch American astronauts into orbit in the MERCURY PROJECT.

Aurora is a display of colored light in the sky near the Earth's north or south pole. Charged particles from the Sun are attracted to the poles. When the particles enter the upper atmosphere, they make the gas there glow. In the northern hemisphere, this effect is called the aurora borealis, or northern lights. In the southern hemisphere, it is known as the aurora australis, or southern lights.

B

Baade, Walter (1893–1960) was a German astronomer who discovered that there are two main populations of stars, identifiable by slightly different chemical compositions. The older generation stars are found at the centers of galaxies, while stars of the younger generation, which include the Sun, are found in the spiral arms of galaxies.

Background radiation is a slight warmth in the Universe, detectable at short radio wavelengths, which is believed to be energy left over from the BIG BANG. According to measurements of the background radiation, space is not completely cold but has a temperature of 2.7 degrees above absolute zero.

Baikonur – see TYURATAM

Barnard, Edward Emerson (1857–1923) was an American astronomer who discovered numerous comets and the fifth moon of Jupiter. He made a famous series of photographs of the Milky Way. In 1916 he discovered BARNARD'S STAR.

Barnard's star is a red dwarf, and the second closest star to the Sun. It lies 6 light-years away in the constellation Ophiuchus, the serpent bearer, but is too faint to be seen without a telescope. Barnard's star is believed to have planets.

Barred spiral galaxy is a galaxy in which the

stars and gas near the center are arranged into a long, straight bar. Spiral arms curve from the ends of the bar.

Bessel, Friedrich Wilhelm (1784–1846) was a German astronomer who in 1838 made the first measurement of the distance of a star, 61 Cygni. He measured the star's distance from its PARALLAX. Bessel also deduced the existence of the white dwarf companions of the stars PROCYON and SIRIUS.

Betelgeuse is a red giant star, about 650 light-years away in the constellation ORION. Betelgeuse varies in size between about 300 and 500 times the diameter of the Sun.

Big Bang is the giant explosion which is believed to have marked the origin of the Universe as we know it. The Big Bang is estimated to have occurred about 18,000 million years ago, and the Universe has been expanding ever since.

Binary star is a pair of stars orbiting around a common center of gravity. Some binaries can be seen as double in a telescope, but others are so close together that they can only be identified as binaries by analysis of their light; these are known as spectro-scopic binaries. In some binaries, such as ALGOL, the stars eclipse each other; these are known as eclipsing binaries.

Black hole is an area of space in which the pull of gravity is so strong that nothing can escape, not even light. Black holes are believed to form when giant stars collapse at the ends of their lives. Giant black holes may exist at the centers of galaxies, where they provide the power source for objects such as QUASARS.

Brahe, Tycho (1546–1601) was a Danish astronomer who made the most accurate observations of the planets in the days before the telescope was invented. He believed that the Sun orbited the Earth, but that the other planets orbited the Sun. Johannes KEPLER used Tycho's observations to work out the true motions of the planets.

von Braun, Wernher (1912–1977) was a German-American rocket engineer who, in wartime Germany, designed the V2 rocket and later, in the United States, designed the SATURN family of rockets that took men to the Moon.

C

Callisto is Jupiter's second largest satellite, 2,995 miles in diameter. It orbits Jupiter every 16.69 days at an average distance of 1 mile.

Cancer, the crab, is a faint constellation of the northern sky. The Sun passes through it from late July to mid-August.

Canopus is the second brightest star in the sky. It lies about 110 light-years away in the southern hemisphere constellation of Carina, the keel, and is approximately 25 times the diameter of the Sun.

Cape Canaveral, in Florida, is the main site used by the United States for space launchings. It was temporarily renamed Cape Kennedy from 1963 to 1973.

Capella is a bright yellow star, 45 light-

Astronauts from the U.S. Apollo project carry out experiments on the surface of the Moon; the second astronaut is clearly visible in the reflection in the visor.

years away in the constellation Auriga, the charioteer.

Capricornus, the sea goat, is a constellation of the southern hemisphere of the sky. The Sun passes in front of it in late January to mid-February.

Cassegrain telescope is a popular design of reflecting telescope, named after the French physicist N. Cassegrain who invented it in 1672. Light collected by the main mirror is reflected on to a secondary, which reflects it back to an eyepiece mounted in a hole in the middle of the main mirror.

Cassini, Jean Dominique (1625–1712) was a French astronomer, born in Italy, who discovered four satellites of Saturn, and drew attention to the gap in Saturn's rings now known as Cassini's division.

Cassiopeia is a famous constellation of the northern sky, representing a queen of Greek mythology. Cassiopeia has a distinctive W shape.

Castor is a bright white star 45 light-years away in the constellation GEMINI. Castor is actually a system of six stars linked by gravity.

Centaurus, the Centaur, is a constellation in the southern hemisphere of the sky, containing the closest star to the Sun, PROXIMA CENTAURI.

Cepheid variable is a type of star that expands and contracts in size, changing in brightness as it does so. The period of variability of a Cepheid is directly related to its average brightness, the brightest Cepheids taking the longest to vary. Since it is so easy to measure the brightness of a Cepheid by observing its variation period,

astronomers use Cepheids as important distance indicators.

Ceres is the largest asteroid, 62 miles in diameter, and the first to be discovered, by Giuseppe PIAZZI in 1801. Ceres orbits the Sun every 4.6 years between the orbits of Mars and Jupiter.

Chromosphere is a layer of hot gas about 9,940 miles deep surrounding the visible surface of the Sun (the PHOTOSPHERE). The chromosphere is only visible in special instruments or at total solar eclipses when the Moon blocks out the dazzling light from the photosphere. The chromosphere is then visible as a red strip of light, from which it takes its name meaning "color sphere."

Comet is a body consisting of rocky blocks and dust particles cemented into a "dirty snowball" by frozen gas. Comets orbit the Sun on elongated paths. When closest to the Sun, the gases of the comet melt to form the flowing tail. Dust released from comets burns up in the atmosphere to produce METEORS.

Conjunction is an alignment of celestial bodies, when a planet lies in the same direction as the Sun. In the case of Mercury and Venus, which can come between the Sun and Earth, astronomers distinguish between INFERIOR CONJUNCTION, when

127

they are on the near side of the Sun, and SUPERIOR CONJUNCTION, on the far side of the Sun.

Constellation is a pattern of stars in the sky. A total of 88 constellations covers the entire sky, many of them representing figures from ancient mythology.

Copernicus, Nicolaus (1473–1543) was a Polish astronomer who proposed that the Earth orbited the Sun like a normal planet, rather than being the center of the universe as had been previously assumed. This view was confirmed by the observations of GALILEO and the calculations of Johannes KEPLER.

Corona is the thin gaseous atmosphere of the Sun, visible as a pearly glow around the Sun at a total eclipse. It consists of gas boiled off from the Sun's surface.

Crab nebula, 6,000 light-years away in TAURUS, is the remains of a star which was seen by oriental astronomers to flare up as a SUPERNOVA in AD 1054.

Crater is a roughly circular depression on the surface of a planetary body, usually caused by the impact of a smaller solid projectile.

Crux, the southern cross, is the smallest constellation in the sky.

Cygnus, the swan, is a constellation of the northern sky often called the northern cross because of its shape. Its brightest star is DENEB.

D

Declination is a coordinate for locating objects in the sky, the celestial equivalent of latitude.

Deimos is the smaller and more distant moon of Mars, roughly 8 miles in diameter, orbiting Mars every 1.26 days at an average distance of 12,427 miles from the planet's center.

Deneb is a white star 1,500 light-years away in the constellation CYGNUS.

Doppler effect is the change in wavelength of light from an object, caused by the object's motion. If the object is receding, then its light is lengthened (reddened) in wavelength. The RED SHIFT of light from galaxies shows that the universe is expanding.

E

Eclipse occurs when the Earth and Moon enter each other's shadow. When the Moon passes in front of the Sun, a solar eclipse is seen in those areas on which the Moon's shadow falls. When the Moon enters the Earth's shadow, a lunar eclipse ensues. On average, one or two solar and lunar eclipses are visible from a given place each year.

Eclipsing binary – see BINARY STAR.

Ecliptic is the path followed by the Sun around the sky each year. It gets its name because only when the Sun and Moon are both on or near this line can eclipses occur.

Effelsberg Radio Observatory, near Bonn, West Germany, is the site of the world's largest fully steerable radio dish, 328 feet in diameter.

Electron is an atomic particle with a negative electric charge. In an atom, the rapidly moving electrons orbit in 'shells' around the nucleus.

Elements, chemical. The most abundant element in the Universe is hydrogen, comprising about 90 percent of all matter. Helium accounts for another 10 percent. All other elements, which astronomers term the heavy elements, make up only about 1 percent of the Universe. Originally, the Universe is believed to have contained only hydrogen and helium, formed in the BIG BANG. The heavier elements have been built up by nuclear reactions inside stars, and distributed through space by SUPERNOVAE.

Elliptical galaxy is a galaxy made of old stars, with no spiral arms. Elliptical galaxies range in size from almost globular to a cross-sectional shape like that of a football.

Encke's comet is the comet of shortest known orbital period, 3.3 years. The comet was discovered in 1818 by the French astronomer Jean Louis Pons, and had its orbit calculated by the German astronomer Johann Franz Encke (1791–1865).

Equatorial mounting is a form of mounting for telescopes. One axis of the mount is aligned parallel with the axis of the Earth; by turning this axis, the rotation of the Earth can be counteracted, so that the image stays still in the telescope's field of view.

Equinox is the instant when the Sun crosses the celestial equator. At the equinoxes, which occur in late March and September, the Sun is overhead on the equator at noon, and everywhere on Earth has equal day and night; the name *equinox* means "equal night."

Europa is the smallest of the four main moons of Jupiter, 1,942 miles in diameter. It orbits Jupiter every 3.55 days at an average distance of about 416,900 miles.

European Space Agency is an organization of European countries involved in space research. ESA is involved in satellite projects both on its own and in conjunction with NASA.

Explorer satellites are a continuing series of American scientific satellites. The first American satellite was Explorer 1, launched on January 31, 1958.

F

Flare is a brilliant outburst on the surface of the Sun caused by a local eruption of energy, usually near a SUNSPOT. Flares eject atomic particles into space which cause radio interference on Earth.

G

Gagarin, Yuri (1934–1968), a Soviet cosmonaut, was the first man to fly in space. On April 12, 1961, he orbited the Earth once in his spaceship Vostok 1.

Galaxy is a collection of stars bound together by gravity. The smallest galaxies contain about a million stars, whereas the largest contain a million times more. Most galaxies are spiral in shape, like our own

MILKY WAY, but about one in five is an ELLIPTICAL GALAXY.

Galileo Galilei (1564–1642) was an Italian scientist who made the first serious astronomical observations with a telescope. He found that Venus shows phases, so that it must orbit the Sun, and he also discovered the four brightest moons of Jupiter. With his telescope, Galileo saw that there were countless stars invisible to the naked eye. His observations helped establish the theory of COPERNICUS that the Earth orbits the Sun.

Galle, Johann Gottfried (1812–1910) was a German astronomer who on September 23, 1846, discovered the planet NEPTUNE close to the position calculated by J. C. ADAMS and U. J. J. LEVERRIER.

Ganymede is the largest moon of Jupiter, 3,275 miles in diameter. Ganymede orbits Jupiter every 7.15 days at an average distance of just over 621,000 miles.

Gemini, the twins, is a famous constellation of the northern hemisphere of the sky. Its two brightest stars are POLLUX and CASTOR. The Sun passes in front of Gemini in late June and July.

Gemini project was a series of American space flights in the two-man Gemini capsule. During the Gemini program, in 1965 and 1966, astronauts learned the techniques of rendezvous and docking of spacecraft and space walks that were vital to the APOLLO PROJECT.

Geosynchronous orbit is an orbit used by artificial satellites, particularly communications satellites, 22,307 miles above the equator. At this height the satellite moves at the same rate as the Earth spins, and thus appears to hang stationary over a point of the equator.

Glenn, John (b.1921) was the first American to orbit the Earth. He performed three orbits in his Mercury capsule on February 20, 1962.

Globular cluster is a ball-shaped group of stars found around the center of a galaxy. There are about 125 globular clusters arranged in a halo around our own galaxy, each containing between about 100,000 and a million stars.

Goddard, Robert H. (1882–1945) was an American rocket pioneer who on March 19, 1926, launched the world's first liquid-fueled rocket.

Gravity is the general force of attraction between bodies in the universe. It does not operate within atoms, whose particles are controlled by nuclear forces.

Great Bear – see URSA MAJOR

H

Hale, George Ellery (1868–1938) was an American astronomer who discovered that sunspots are cooler areas on the Sun associated with magnetic fields. Hale was the founder of three major observatories: YERKES observatory, and the observatories on Mount Wilson and Mount Palomar (now called HALE OBSERVATORIES).

Hale Observatories is the name given since 1970 to the astronomical observatories on Mount Wilson and Mount Palomar in

A cluster of galaxies in the constellation Hercules, taken through the 200-inch Hale telescope.

California, both founded by George HALE. Mount Wilson contains the famous 8-foot reflector opened in 1917. On Mount Palomar is the 16-foot Hale reflector, which was the world's largest until surpassed by a larger telescope at the ZELENCHUKSKAYA OBSERVATORY, USSR.

Halley, Edmond (1656–1742) was a British astronomer who in 1705 calculated the orbit of the comet that bears his name. In 1720 he became second Astronomer Royal, and spent the rest of his life at Greenwich observing the complex motions of the Moon.

Halley's comet is a famous comet that orbits the Sun every 76 years. Its next return is due in 1986, although it will not be easy to see.

Hercules is a major constellation of the northern hemisphere of the sky, named after a hero from Greek mythology.

Herschel, Sir John (1792–1871) was an English astronomer, son of Sir William, who made a major survey of the southern skies from the Cape of Good Hope, thereby completing the survey of the heavens begun in the northern hemisphere by his father.

Hershel, Sir William (1738–1822) was an English astronomer, born in Germany, who discovered the planet URANUS. He made a major survey of the northern sky, cataloguing all objects of interest. His son, John, extended the work to the southern hemisphere. Between them, this father-and-son combination were among the greatest observers in the history of astronomy.

Hertzsprung Russell diagram is a graph in which the temperature of stars is plotted against their brightness. A star's position on the H-R diagram shows whether it is a normal star (on the MAIN SEQUENCE), a red giant (large but cool) or a white dwarf (small but hot).

Hipparchus (2nd century BC) was a Greek astronomer who is regarded as the greatest astronomer of antiquity. He made many accurate measurements of the motion of the Earth in space, including the length of the year, and made an important catalogue of 850 stars in which he introduced the MAGNITUDE system.

Hoyle, Sir Fred (b. 1915) is an English astronomer best known for his support of the STEADY STATE THEORY of the Universe. Hoyle also proposed that the chemical elements were built up from hydrogen by nuclear reactions inside stars, an idea now accepted.

Hubble, Edwin (1889–1953) was an American astronomer who in the 1920s discovered that galaxies are far outside our own galaxy, and also found that the Universe is expanding. He made his first discovery by identifying individual stars in the ANDROMEDA GALAXY, and his second discovery by studying the RED SHIFT in light from galaxies.

Hubble constant is a figure that shows how fast the Universe is expanding, and thus how long ago the BIG BANG took place. According to latest measurements, galaxies move at about 10 miles a second for every million light-years they are apart.

Huygens, Christian (1629–1695) was a Dutch scientist who in 1655 discovered Saturn's largest moon, TITAN, and explained that Saturn's rings are made of a swarm of tiny moonlets.

Hyades is a cluster of about 200 stars, 148

light-years away in the constellation TAURUS.

Hydrogen in space. Hydrogen is the simplest and most abundant of all the ELEMENTS in space. Stars are made mostly of hydrogen, as are the glowing clouds of gas such as the ORION NEBULA. Hydrogen emits radio waves at 21-centimeters (see TWENTY-ONE CENTIMETER RADIATION).

I

Inclination is the angle of a planet's orbit to the Earth's orbit, or of a satellite to the Earth's equator.

Inferior conjunction is the instant when Mercury or Venus are in line between the Earth and Sun.

Infra-red astronomy is the study of the Universe at wavelengths longer than those of red light. These studies reveal objects such as gas clouds and forming stars which are too cool to emit visible light, and brilliant objects such as the centers of galaxies that are otherwise obscured by dust.

Intelsat is the name of the communications satellites that provide a global telephone, radio, and TV network. Early Bird, the first, was launched in 1965. The current Intelsat IV A series can each carry 6,000 telephone calls.

Interstellar molecules are molecules in space, usually detected by the radio waves they emit. Over 30 interstellar molecules of varying complexity are known, including water, ammonia and formaldehyde. Their presence indicates the existence of dense gas clouds, such as those from which stars form.

Io is the nearest of the four main satellites of Jupiter, orbiting every 1.77 days at an average distance of 262,218 miles. Io is 2,257 miles in diameter.

J

Jansky, Karl (1905–1950) was an American radio engineer who in 1932 discovered radio waves coming from the Galaxy, thereby founding the study of radio astronomy.

Jodrell bank is the location near Macclesfield, Cheshire, of the radio observatory of the University of Manchester. Its main instrument is the famous 249-foot diameter dish.

Juno was the third known asteroid, discovered in 1807. It is about 143 miles in diameter and orbits the Sun every 4.36 years between the orbits of Mars and Jupiter.

Jupiter is the largest planet of the solar system. It is a large ball of gas, similar in composition to the Sun, but may have a central rocky core. The visible surface is not solid, but consists of swirling, multi-colored clouds, the only permanent feature of which is the red spot, apparently the top of a storm cloud.

K

Kepler, Johannes (1571–1630) was a German mathematician and astronomer who worked out the laws of planetary motion (KEPLER'S LAWS) from the observations of Tycho BRAHE. Kepler's work finally established that the Earth is a planet orbiting the Sun, as COPERNICUS had proposed.

Kepler's laws are the three laws of planetary motion calculated by Johannes KEPLER. The first, and most important law, is that planets orbit the Sun in elliptical paths (not circular paths as had been assumed previously). Secondly, the planet moves fastest when it is nearest to the Sun. These two laws were published in 1609. The third law, published in 1619, notes that there is a direct connection between a planet's orbital period and its distance from the Sun.

Kitt Peak Observatory is an astronomical observatory near Tucson, Arizona, containing the largest collection of telescopes in the world. Its main telescope is a 13-foot reflector.

L

Landsat is the name of two American satellites for surveying the Earth. Their photographs are used to make maps of remote countries, identify locations of new mineral resources, detect areas of crop disease and monitor pollution.

Leavitt, Henrietta (1868–1921) was an American astronomer who discovered the relationship between the period of variability of a CEPHEID VARIABLE and the star's average brightness, a discovery which was used in measuring the size of the galaxy by Harlow SHAPLEY.

Lemaitre, Georges (1894–1966) was a Belgian astronomer who originated the BIG BANG theory of cosmology.

Leo, the lion, is a constellation of the equatorial region of the sky. The Sun passes in front of Leo from mid-August to mid-September. The brightest star in Leo is REGULUS.

Leonov, Alexei (b. 1934), a Soviet cosmonaut, was the first man to walk in space. He spent 10 minutes outside the Voskhod 2 spacecraft on March 18, 1965.

Leverrier, Urbain (1811–1877) was a French astronomer who, like the Englishman J. C. ADAMS, calculated the existence of the planet Neptune. He sent his results to the Berlin observatory, where the planet was discovered by J. G. GALLE.

Libra, the scales, is a faint constellation of the southern hemisphere of the sky. The Sun passes in front of Libra during November.

Lick Observatory is the astronomical observatory of the University of California, on Mount Hamilton. Its main telescopes are a 10-foot reflector and a 36-inch refractor.

Light-year is the distance traveled by a beam of light in one year. It is equivalent to 6 trillion miles.

Local Group is the cluster of about 20 known galaxies of which our own galaxy is the second largest member. The largest member of the local group is the ANDROMEDA GALAXY.

Lowell Observatory is an astronomical observatory at Flagstaff, Arizona, founded in 1894 by Percival LOWELL. Its main telescope is a 6-foot reflector.

Lowell, Percival (1855–1916) was an American astronomer who believed in the existence of canals on Mars dug by intelligent beings. He also predicted the existence of another planet beyond Neptune, and initiated a search at his own observatory which culminated in the discovery of Pluto by Clyde TOMBAUGH.

Luna probes are a series of Soviet Moon probes, members of which have photographed the Moon and landed on its

Meteors appear in the sky as a shower of streaks.

surface. Two automatic Moon rovers have been delivered by Lunar probes, and others have automatically brought lunar samples back to Earth.

Lunik spacecraft launched from the USSR in 1959 were the first probes to strike the Moon and photograph the side which is turned away from Earth.

Lyra, the lyre, is a small but prominent constellation of the northern sky. Its brightest star is VEGA.

M

Magellanic clouds are two satellite galaxies of our own Milky Way. They are each about 160,000 light-years away, and are about 1/30 and 1/200 the size of the Milky Way.

Magnitude is the scale that measures a star's apparent brightness as seen from Earth. The faintest stars visible to the naked eye are called magnitude 6; they are 100 times fainter than the first magnitude stars. Objects even brighter are given negative (minus) magnitudes. Objects fainter than magnitude 6 are given progressively larger positive magnitudes.

Main sequence is the classification given to stars in the healthy prime of their lives, when they are burning hydrogen at their centers to create energy, as is the Sun. Main-sequence stars form a band running across the HERTSPRUNG-RUSSEL DIAGRAM.

Mare (plural **maria**) is the name given to the large, dark plains which extend over much of the Earth-turned hemisphere of the Moon.

Mariner spacecraft are a series of American planetary probes. Mariner 9 made a complete photographic map of Mars in 1971 and 1972, and Mariner 10 took the first closeup photographs of Mercury and Venus in 1974.

Mars is the fourth planet in line from the Sun. It is known as the red planet, because of the distinctive color of its surface rocks, caused by extensive amounts of iron oxide. There is little air or water on Mars, and temperatures are frigid. Despite searches by the VIKING PROBES, there is no sign of life on Mars.

McDonald Observatory is an astronomical observatory near Fort Davis, Texas with 6.5-foot and 8.8-foot reflectors.

Mercury is the nearest planet to the Sun, not much larger than our own Moon. Its surface is lunar-like, covered with craters presumably formed by meteorite impacts. Mercury has no real atmosphere.

Mercury project was the first American attempt to fly men in space, in the one-man Mercury capsule. The program lasted from May 1961 to May 1963, during which time a total of 6 astronauts were launched into space, the longest flight lasting 34 hours.

Messier, Charles (1730–1817) was a French astronomer who made a list of over 100 nebulae and clusters of stars, many of which are still known by their Messier or M numbers.

Meteor is a dust particle from space, seen as it burns up in the atmosphere. Meteors are believed to be dust from comets. When the

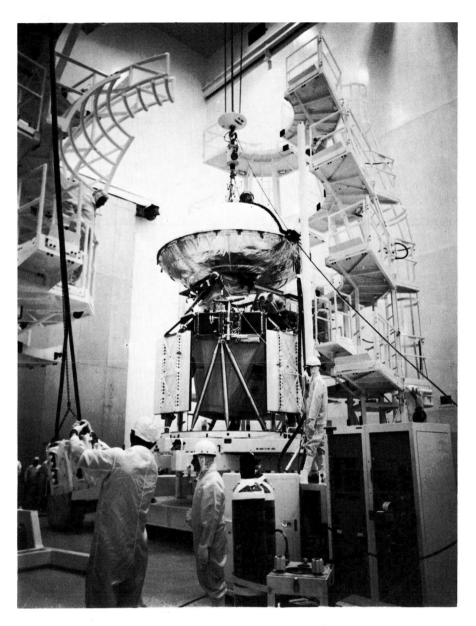

Earth crosses the orbit of a comet, as happens several times a year, a meteor shower is seen. Most meteors are about the size of a grain of sand, and burn up long before reaching the Earth's surface.

Meteorite is a lump of rock or metal from space that penetrates the atmosphere to reach the Earth's surface. If a meteorite is large enough it will blast a crater, like the one nearly a mile wide in Arizona. Otherwise it will fall harmlessly to the ground like the world's largest known meteorite, weighing 60 tons, near Grootfontein in Namibia (southwest Africa).

Milky Way is the faint band of starlight seen crossing the sky on clear, dark evenings. The Milky Way is actually the lane of our galaxy seen from inside, therefore our galaxy is also known as the Milky Way. There are about 100,000 million stars in the Milky Way galaxy, which is 100,000 light-years in diameter.

Minor planet – see ASTEROID.

Moon is the Earth's nearest natural neighbor in space. It is a rocky body pitted with countless craters believed to have been formed by meteorite impacts. In some

Inside the launch facility of the U.S. National Aeronautics and Space Administration, usually known as NASA, the first Viking Orbiter and "nuclear-powered" lander are mated.

places, volcanic eruptions have produced the dark lava plains known as *maria* ("seas"). The Moon turns on its axis in the same time as it takes to orbit the Earth, so that it keeps one face permanently turned towards us. The Moon is airless, waterless and lifeless.

Mount Wilson and Palomar Observatories – see HALE OBSERVATORIES.

Mullard Radio Observatory, at Cambridge, England, is the site of two large aperture synthesis radio telescopes, the One Mile and the Five Kilometres instruments (the names signify their length). PULSARS were discovered at the observatory, which is operated by the University of Cambridge.

N

NASA, the National Aeronautics and Space Administration, is the American government agency for civilian space flight, founded in 1958.

Nebula (plural **nebulae**) is a mass of gas and dust in space. Some are believed to be the sites of star formation, such as the ORION NEBULA. Some nebulae are made to glow by the light of stars inside them, others shine by reflection, but others are dark and invisible.

Neptune is the eighth planet in average distance from the Sun, discovered in 1846 by J. G. GALLE. Neptune has a deep gaseous atmosphere, underneath which is believed to be a core of rock and ice. The planet appears as nothing more than a greenish disk in even a large telescope.

Neutron star is a small, highly compressed star about 6 to 12 miles in diameter left behind after the death of a star more massive than the Sun. In a neutron star, gravity has compressed the electrons and protons of the star's atoms to form the atomic particles called neutrons. See also PULSARS.

Newton, Isaac (1642–1727) was an English scientist whose laws of motion and gravity explained why planets orbit the Sun as deduced by Johannes KEPLER. Newton's work on light led to his design of a reflecting telescope.

NGC is an abbreviation for the New General Catalogue of star clusters and nebulae, compiled in 1888 by the Danish astronomer J. L. E. Dreyer and widely referred to by astronomers.

Nova is a stellar outburst caused by the transfer and ignition of gas between two stars in a binary system. The star that causes the outburst is a WHITE DWARF, on to the surface of which gas flows from a normal companion, and then erupts in a nuclear explosion.

Nutation is a slight nodding of the Earth's axis caused by the uneven gravitational pulls of the Sun and Moon. Nutation slightly alters the Earth's tilt in space every 18.6 years.

O

Object glass is the main lens at the front of a refracting telescope.

Occultation is the covering of a celestial body by the Moon. The occultation of stars by the Moon is used to track the Moon's orbit around the Earth.

Opposition is the instant when a planet is opposite in the sky to the Sun. At opposition, a planet is seen due south at midnight.

Orbit is the path in space of one body around another. Orbits are usually elliptical in shape, although the orbits of the planets scarcely depart from circles.

Orbiter spacecraft, 5 of which were launched from the USA in 1966 and 1967, orbited the Moon at heights of a few hundred miles and took detailed photographs of practically the whole surface, showing details down to $3\frac{1}{2}$ feet across.

Orion is a major constellation in the equatorial region of the sky, representing a figure from Greek mythology. Its two brightest stars are RIGEL and BETELGEUSE, and it also contains the ORION NEBULA.

Orion nebula is a glowing mass of gas about 1,500 light-years away in the constellation ORION. New stars are forming inside the nebula, and it is their light which makes it glow. A whole cluster of stars will probably be formed from the Orion nebula.

Oscillating universe is the theory which says that the current expansion of the Universe will eventually slow, stop, and be reversed, so that the Universe collapses again to another BIG BANG. According to this theory, the Universe continues in endless cycles of expansion and contraction. However, astronomers do not presently believe that the expansion of the Universe will be halted.

Ozone Triatomic oxygen (O_3)

P Q

Pallas is the second largest asteroid, 378 miles in diameter, and the second to be discovered, by Wilhelm Olbers in 1802. Pallas orbits the Sun every 4.6 years between the orbits of Mars and Jupiter.

Parallax is the slight shift in position of a nearby star when viewed from opposite sides of the Earth's orbit. The amount of parallax shift reveals the star's distance. Stars beyond about 100 light-years distance have parallaxes too small to be accurately measured.

Parsec is a unit of distance in astronomy, equivalent to 3.26 LIGHT-YEARS. It is the distance at which a star would show a PARALLAX of 1 second of arc, although no star is actually quite this close.

Pegasus is a large constellation of the northern hemisphere of the sky, representing the winged horse of Greek mythology. Its main feature is a great square marked out by four stars.

Perigee is the nearest point to the Earth that a body such as an artificial satellite reaches in its orbit.

Perihelion is the nearest point to the Sun that a body such as a planet reaches in its orbit.

Perseus is a constellation of the northern sky, representing a figure from Greek mythology. Its most famous star is ALGOL.

Phobos is the larger and nearer of the two moons of Mars, roughly 14 miles in diameter, orbiting every 7.65 hours 3,700 miles from the planet's center.

Photon A "packet" or quantum of light radiation.

Photosphere is the visible surface of the Sun, a layer of gas at a temperature of about 10,830° F.

Piazzi, Giuseppe (1746–1826) was an Italian astronomer who on January 1, 1801, discovered the first asteroid, CERES.

Pioneer probes are a series of American craft to investigate the Solar System. Pioneers 10 and 11 took remarkable close-up photographs of Jupiter in 1973 and 1974.

Pisces, the fishes, is a constellation of the equatorial region of the sky. The Sun passes in front of Pisces from mid-March to mid-April.

Planet is a large non-luminous body in orbit around a star, which can be made of rock or gas. Planets are not large enough to generate energy at their centers by nuclear reactions, as do stars.

Planetary nebula is a shell of gas ejected from a red giant at the end of its life, leaving the star's hot core as a central WHITE DWARF. Planetary nebulae are so named because they superficially resemble a planet's disk when seen through a small telescope.

Pleiades are a group of 200 stars about 400 light-years away in the constellation TAURUS.

Plow see URSA MAJOR.

Pluto is the planet with the greatest average distance from the Sun (over 3,666 million miles), discovered in 1930 by Clyde TOMBAUGH. Actually, its orbit is so eccentric that from 1979 to 1999 it comes closer to the Sun than Neptune. Many astronomers think that Pluto is an escaped satellite of Neptune. Latest measurements suggest that it is the smallest planet of the Solar System.

Polaris, the pole star, lies approximately 400 light-years away in the constellation URSA MINOR.

Pollux is an orange giant star, 36 light-years away in the constellation GEMINI, of which it is the brightest star.

Precession is a wobbling of the Earth on its axis every 26,000 years like a spinning top, caused by the gravitational pulls of the Sun and Moon.

Procyon is the brightest star in the constellation Canis Minor, 11.4 light-years away. It has a WHITE DWARF companion.

Prominences are eruptions of gas from the surface of the Sun, often associated with sunspots.

Proper motion is a change in a star's position over time caused by its movement around the Galaxy. The proper motions of stars are not noticeable to the naked eye, but over thousands of years they slowly change the shapes of the constellations.

Proxima Centauri is a red dwarf star 4.3 light-years away, the closest member of the ALPHA CENTAURI triple system.

Ptolemy (c. 100–c. AD 178) was a Greek scientist who in his book the *Almagest* put forward the Earth-centered view of the Universe that was accepted until the time of COPERNICUS.

Pulsar is a rapidly rotating NEUTRON STAR that gives out a radio pulse every time it spins. Pulsars were discovered by Cambridge radio astronomers in 1967. The fastest pulsar, at the center of the CRAB NEBULA, flashes 30 times a second; the slowest ones flash every 3 seconds or so.

Quasar is a small but brilliant object far off in space. Quasars can give out the energy of hundreds of galaxies from a space not much larger than our own solar system.

R

Radio astronomy is the study of the Universe in radio waves, emitted naturally by many objects in space. Radio astronomy has led to the discovery of objects such as QUASARS and PULSARS and has also helped astronomers map the structure of our own

Galaxy through the TWENTY-ONE CENTI-METER RADIATION of hydrogen.

Radio galaxy is a galaxy that emits considerable radio energy, both from a central point and from lobes either side of the galaxy that seem to have been ejected in explosions. Radio galaxies are closely related to QUASARS.

Radio telescope is a device for collecting radio waves from space. Most radio telescopes are like reflecting telescopes in design, but because radio waves are so much longer than light waves, radio telescopes have to be correspondingly larger to see the sky in as much detail.

Red dwarf is a faint star that is smaller and cooler than the Sun. Because they burn so slowly, red dwarf stars live very long.

Red giant is a large star with a cool surface, which has swelled up towards the end of its life. In about 5,000 million years, our Sun will become a red giant like ARCTURUS.

Red shift is the lengthening in the wave-length of light from a receding object, such as a galaxy. The amount of red shift, which is caused by the DOPPLER EFFECT, reveals how fast the object is receding.

Reflecting telescope is a telescope that uses a mirror to collect and focus light. The first reflector was built in 1668 by Isaac NEWTON, but the idea had previously been proposed in 1663 by the Scottish scientist James Gregory.

Refracting telescope is a telescope that uses lenses to collect and focus light. The invention of the refractor is attributed to the Dutch optician Hans Lippershey in 1608, although he was almost certainly not the first to make one.

Regulus is a blue-white star about 84 light-years away in the constellation LEO, of which it is the brightest star.

Rigel is a blue giant star 850 light-years away, the brightest star in ORION. Rigel is 78 times the Sun's diameter.

Right ascension is a coordinate for locating objects in the sky, the celestial equivalent of longitude.

Rosse, Lord (1800–1867) was an Irish astronomer who in 1845 completed a 6-foot reflector, then the largest in the world, with which he studied nebulae and star clusters.

Royal Greenwich Observatory was set up in 1675. In 1958 it moved from Greenwich to Herstmonceux, Sussex. Its largest telescope is an 8.2-foot reflector.

S

Sagittarius, the archer, is a constellation of the southern hemisphere of the sky. The Sun passes in front of Sagittarius from mid-December to mid-January.

Salyut is a Soviet space station, smaller

This giant radio telescope is at the Anglo-Australian observatory in Siding Spring, Australia. It has a 4-meter reflector.

than the American SKYLAB. The first of the Salyut series was launched in 1971.

Satellites are any bodies that move in an orbit around another, more massive body. The planets, for example, are satellites of the Sun. And the Moon is a satellite of the Earth. Many other artificial, or man-made satellites orbit the Earth, too. They are used for such purposes as radio and television communications, and for gathering information on the Earth and the weather.

Saturn is the sixth planet from the Sun. Like Jupiter, it is a ball of gas. Saturn's most interesting feature is its rings, about 170,000 miles in diameter, made of tiny ice-covered blocks of rock.

Saturn rockets are two related rockets designed for manned launchings. The smaller Saturn 1B was used for launching Apollo capsules into Earth orbit; the larger Saturn V was used to launch the Apollo Moon missions.

Schmidt, Maarten (b. 1929) is an American astronomer who in 1963 measured the RED SHIFT of the first quasar, 3C 273.

Schmidt telescope is a wide-angle photographic telescope using both lenses and mirrors, designed in 1930 by the Estonian optician Bernhard Schmidt.

Scorpius, the scorpion, is a constellation of the southern sky whose brightest star is ANTARES. The Sun passes across part of Scorpius at the end of November.

Seyfert galaxy is a galaxy with a brilliant core, rather like a scaled-down QUASAR. Astronomers think that Seyfert galaxies are closely related to quasars.

Shapley, Harlow (1885–1972) was an American astronomer who first measured the size of our Milky Way galaxy, and showed that the Sun lies about two-thirds of the way to the edge.

Shepard, Alan (b. 1923) was the first American astronaut to fly in the MERCURY PROJECT. He did not go into orbit, but completed a brief suborbital flight. In 1971 he commanded the Apollo 14 Moon-landing mission.

Shooting Star see METEOR.

Sidereal period is the time taken for an object such as a planet or satellite to complete one orbit relative to the stars.

Sirius is the brightest star in the sky, 8.7

Skylab 1 astronauts took this photograph of a massive explosion of the Sun, when light and radiation spewed into space from a solar flare.

light-years away in the constellation Canis Major. It has a WHITE DWARF companion.

Skylab was an American space station made from a converted Saturn V upper stage. Astronauts spent up to 84 days in Skylab in 1973–1974.

Solar day is the time taken for an object to complete one orbit relative to the Sun.

Solar system is the collection of planets and other objects orbiting the Sun.

Solar transit is the passage of a celestial body, such as the planet Mercury, across the face of the Sun.

Solar wind is a steady stream of electrically charged subatomic particles, given out by the Sun. In Earth's atmosphere, it give rise to the glowing displays called AURORAS.

Solstices are the two periods of the astronomical year when the Sun is farthest north or south of the equator: roughly June and December 21st. They correspond to the longest and shortest day, respectively, in the northern hemisphere and the opposite in the southern hemisphere.

Southern Cross – see CRUX.

Soyuz is a Soviet spacecraft capable of holding two or three men. Soyuz capsules can fly on their own, or are used to ferry cosmonauts to SALYUT space stations.

Space Age is the name often given to the age of scientific discovery that began with the first SPUTNIK in 1957.

Space shuttle is the latest major development of the Space Age, a giant vehicle that

is rocketed into orbit, then glides back to Earth to be fitted up for the next space journey. The USA launched its first space shuttles in 1981. They are designed to carry large loads, including satellites, and to cut launch costs by over 50 percent.

Spacelab is a scientific space station designed in Europe to be carried in orbit by the space shuttle. It made its first successful flight in November/December 1983. On later flights scientists will be able to work in Spacelab for up to a month.

Spectrum is the band of colors obtained when the light from a body is split into its various wavelengths. This is usually done by a spectrograph, which photographs the spectrum. A spectrum gives information about the temperature of a body and the materials composing it.

Spica is a blue-white star 260 light-years away in the constellation VIRGO.

Spiral galaxy is the most common type of galaxy in space. Old stars are arranged in a central bulge, while newer stars form spiral arms around it. Our MILKY WAY galaxy is a spiral, as is the ANDROMEDA GALAXY.

Sputnik was a series of early Soviet satellites. Sputnik 1, launched on October 4, 1957, was the first artificial satellite.

Star is a glowing ball of gas in space. Any quantity of matter above a certain mass will become a star, as it is heated up by thermonuclear reaction. In our own SOLAR SYSTEM, this critical mass can be measured by Jupiter, the Sun's largest planet, which is just too lightweight to have become a star.

Steady state theory, put forward by Fred Hoyle and Thomas Gold, supposes the Universe to have no sudden beginning or end in time, and matter to be continuously created to fill the spaces created by its expansion. It is a cosmology (philosophy of the Universe as a whole) at present out of favor.

Stellar spectroscopy is the analysis of light from stars. It reveals the star's composition, its temperature, and whether it is a dwarf or a giant.

Stonehenge is a monument on Salisbury Plain, England, which some astronomers believe embodies advanced astronomical knowledge. The oldest parts of it date back 4,500 years.

Sun is our nearest star, almost a million miles in diameter. It is believed to have been born, along with the rest of the solar system, 4,600 million years ago. It keeps hot by nuclear reactions at its center which turn the hydrogen of which it is mostly made into helium.

Sunspot is an area on the Sun's surface about 2,700° F cooler than its surroundings, so that it appears darker by contrast. Sunspots are believed to be caused by strong magnetic fields which block the outward flow of heat from the Sun's interior.

Supergiant stars are those 300 or more times as wide as the Sun. Examples are ANTARES and BETELGEUSE.

Superior conjunction is the instant when Mercury and Venus are on the far side of the Sun from Earth.

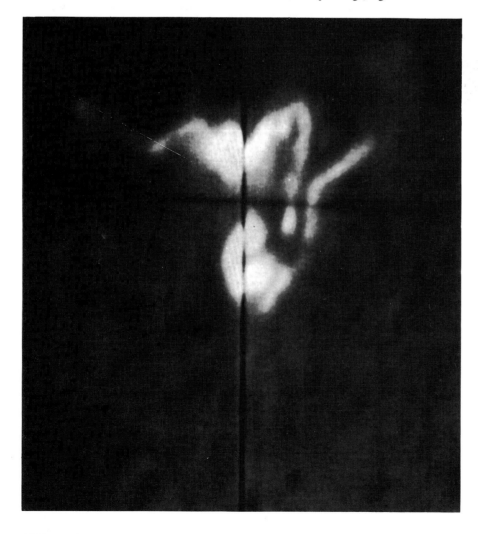

A Viking probe is launched on its way to Mars, carried by a Titan rocket.

Supernova is a tremendous flare-up following the collapse of an aged, giant star. The end product of a supernova is either a tiny NEUTRON STAR or a BLACK HOLE. The Crab nebula may have been caused by a supernova.

Surveyor space probes were launched from the USA from 1966 to 1968 to map and land on the Moon, in preparation for the first manned landings in the Apollo spacecraft.

Synodic period is the time taken for an object to come back to the same position as seen from Earth.

T

Taurus, the bull, is a major constellation of the equatorial region of the sky. Its brightest star is ALDEBARAN, and it also contains the CRAB NEBULA, the HYADES, and the PLEIADES. The Sun passes in front of Taurus from mid-May to late June.

Telescope is a device for collecting and focusing light, thereby revealing faint objects and fine detail otherwise invisible to the naked eye. The most important statistic about a telescope is its aperture, rather than its magnification.

Tereshkova, Valentina (b. 1937) was the first woman to fly in space. She orbited the Earth for nearly three days in the Soviet Vostok 6 spacecraft during June 1963.

Titan is the largest satellite of Saturn, 3,623 miles in diameter. It orbits Saturn every 15.95 days at a distance of 0.75 million miles. It is the only satellite known to have a substantial atmosphere

Titan rocket is an American space launcher used in various designs to orbit GEMINI astronauts, Earth satellites, and planetary probes.

Tombaugh, Clyde (b. 1906) is an American astronomer who discovered the planet PLUTO in 1930 at the LOWELL OBSERVATORY.

Tracking station is an array of telescopes, radio telescopes or cameras that moves to keep track of an artificial satellite, such as a weather satellite.

Transit is the passage of a solar system body across the face of the Sun, or of a celestial object across an observer's north-south meridian.

Triton is the largest moon of NEPTUNE, and may also be the largest moon in the Solar System; its estimated diameter is 3,728 miles. Triton orbits Neptune every 5.88 days at a distance of 220,586 miles.

Tsiolkovsky, Konstantin (1857–1935) was a Soviet prophet of astronautics who at the turn of the century worked out the theory behind rocket propulsion, invented the idea of step rockets, and predicted the establishment of space stations in orbit.

Twenty-one centimeter radiation is emitted naturally by hydrogen gas in space. It is detected by radio astronomers who have used it to trace the spiral shape of our Milky Way galaxy.

Tyuratam is the main Soviet launch site, their equivalent of Cape Canaveral, northeast of the Aral Sea.

U

UFO is short for Unidentified Flying Object, sightings of which are often connected with "visitors from space."

Ultraviolet astronomy is the study of the Universe at wavelengths shorter than those of visible light. Ultraviolet radiation is emitted strongly by very hot stars and gas and thus provides information on energetic processes in the Universe.

Ultraviolet waves are electromagnetic waves shorter and more penetrating than those of visible light but longer than X-rays or gamma rays. Special telescopes can detect UV waves reaching Earth from space from very hot stars and other bodies.

Universe is space plus all its galaxies and other contents, taken as a whole. According to the Theory of Relativity, because space-time curves in the presence of matter, the Universe is "finite but unbounded." That is, it has no end, but a ray of light would travel around it until the light returned to its source.

Uranus is the third largest of the Sun's family of nine planets, and is seventh most distant from its parent star. It is a gas giant, visible through telescopes as a greenish disk. Its axis has the greatest tilt towards the Sun of any planet. It has five known moons and, as recently discovered, a thin orbiting rocky ring.

Ursa Major, the great bear, is a famous constellation in the northern sky. Seven of its stars make up the familiar saucepan shape, also known as the Big Dipper or the Plow.

Ursa Minor, the lesser bear, is a constellation at the north pole of the sky. Its brightest star is POLARIS.

V

Variable stars change in brightness because of variations in the size of the star itself, as in a CEPHEID VARIABLE, because one star in a binary eclipses the other, as with ALGOL, or even because of gas passing between two close stars, as in a NOVA. Over 25,000 variable stars of all types are known.

Vega is a white star, 26 light-years away, in the constellation LYRA.

Venus is the second planet from the Sun, and the brightest object as seen from Earth after the Sun and Moon. The brightness of Venus is partly caused by its unbroken layer of white clouds. Below the clouds, its dense atmosphere of carbon dioxide is too hot for life, producing temperatures up to $896°F$. For reasons unknown; Venus rotates back to front (east to west) very slowly.

Vernal equinox is the moment when the Sun moves into the northern hemisphere of the sky, on or around March 21 each year.

Vesta is the third largest asteroid, 334 miles in diameter, and the fourth to be discovered, by Wilhelm Olbers in 1807. Vesta orbits the Sun every 3.6 years between the orbits of Mars and Jupiter.

Viking probes were two American spacecraft sent to look for life on Mars in 1976. Though both landed successfully, no sure signs of Martian life were detected.

Virgo, the virgin, is a constellation of the

135

equatorial region of the sky, whose brightest star is SPICA. The Sun passes in front of Virgo from mid-September to early November.

Voskhod Soviet spacecraft were developed from the earlier VOSTOK spacecraft to hold a crew of two or three astronauts.

Vostok was a Soviet one-man capsule, a sphere 7.5 feet in diameter, in which Yuri GAGARIN made the first manned spaceflight. A total of six cosmonauts made flights in Vostoks, the longest lasting five days.

Voyager probes are the two latest from the USA to be sent on journeys through and beyond the SOLAR SYSTEM. Launched in 1977, Voyager 1 and 2 have already sent back fascinating details of Jupiter and Saturn, and are now on their way to the outermost planets.

W

Weightlessness is a condition of matter when outside a gravitational field. Astronauts have to live in the condition of weightlessness when they are in orbit or in more distant spaceflight.

White dwarf is an extremely compact, dense star, only about the size of Earth but with the mass of the Sun. A white dwarf is a late stage in the life history of an average star such as the Sun, at a point where the star is using up the last of its fuel.

X

X-rays are electromagnetic waves shorter and more penetrating than those of visible and ultraviolet light, but longer and "softer" than gamma rays. They are emitted by many stars and other celestial objects, and are found with special telescopes.

Y

Yerkes Observatory is the astronomical observatory of the University of Chicago, at Williams Bay, Wisconsin. Its major instrument is a 3.3-foot refractor.

Z

Zelenchukskaya telescope in the Caucasus mountains of western USSR is the world's largest reflecting telescope, with a glass mirror 19.6 feet in diameter. See HALE TELESCOPE.

Zodiac is a region of the night sky that corresponds to the Sun's pathway, containing 12 zodiacal constellations.

A Voyager spacecraft aims its instrument scan platform at the planet Jupiter in this artist's impression of the space probe's journey past Jupiter and continuing on to Saturn.

INDEX

138

Acknowledgments

Pages: 8 Nasa/Science Photo Library; 9 Frank Spooner; 12–13 Hatfield Polytechnic Observatory; 14 British Museum; 15 top Michael Holford, bottom Department of the Environment; 17 Istanbul University; 18 top RTHPL, bottom Mansell; 19 Mansell; 20 RTHPL; 21 Martin-Marietta Aerospace; 22 Nasa; 24 top & bottom left California Institute of Technology, right Lick Observatory; 25 Zefa; 27 Zefa; 29 Science Museum; 30 Hale Observatory; 31 Mount Stromlo Observatory; 32 Mullard Radio Astronomy Observatory; 33 top Max Planck Institute for Radio Astronomy, bottom Arecibo Ionospheric Laboratory; 42 Zefa; 44 Science Photo Library; 45 Sonia Halliday; 46 top Zefa/Photri, bottom and 47 Lockheeds Solar Observatory; 49 Nasa; 51 Lockheeds Solar Observatory, bottom Zefa; 53 Nasa; 54 & 55 California Institute of Technology; 56 Royal Greenwich Observatory; 57, 58 Nasa; 59 Science Photo Library; 61, 63 Nasa; 66 Mansell; 67 top California Institute of Technology, bottom U.S. Naval Observatory; 68 top American Meteorite Laboratory, bottom California Institute of Technology; 69 American Meteorite Laboratory; 70 California Institute of Technology; 73 Nasa; 74 California Institute of Technology; 76 U.S. Naval Observatory; 74 California Institute of Technology; 78/79 California Institute of Technology, insert top Kitt Peak National Observatory; insert bottom California Institute of Technology; 80 Robin Kerrod; 81 & 83 California Institute of Technology; 85 Nasa; 86 & 87 Novosti; 88, 89, 91, 92, 93 Nasa; 98 bottom Weapons Research, Salisbury (Aust.); 99 The Post Office; 100 Nasa; 101 Met. Office; 102 Laserscan; 104, 105 Nasa; 106 Photri/Zefa; 107, 108, 109 Nasa; 110 Photri/Zefa; 111, 112, 113 Nasa; 114 Novosti; 115 & 116 Nasa; 117 Frank Spooner; 121 Nasa; 122 Kerrod/Nasa; 125 Nasa; 127 Nasa; 129 Mount Wilson and Palomar Observatories; 130 Kit Peak; 131 Nasa; 133 Anglo-Australian Observatory; 134 Associated Press; 135 Nasa; 136 Photri/Zefa.

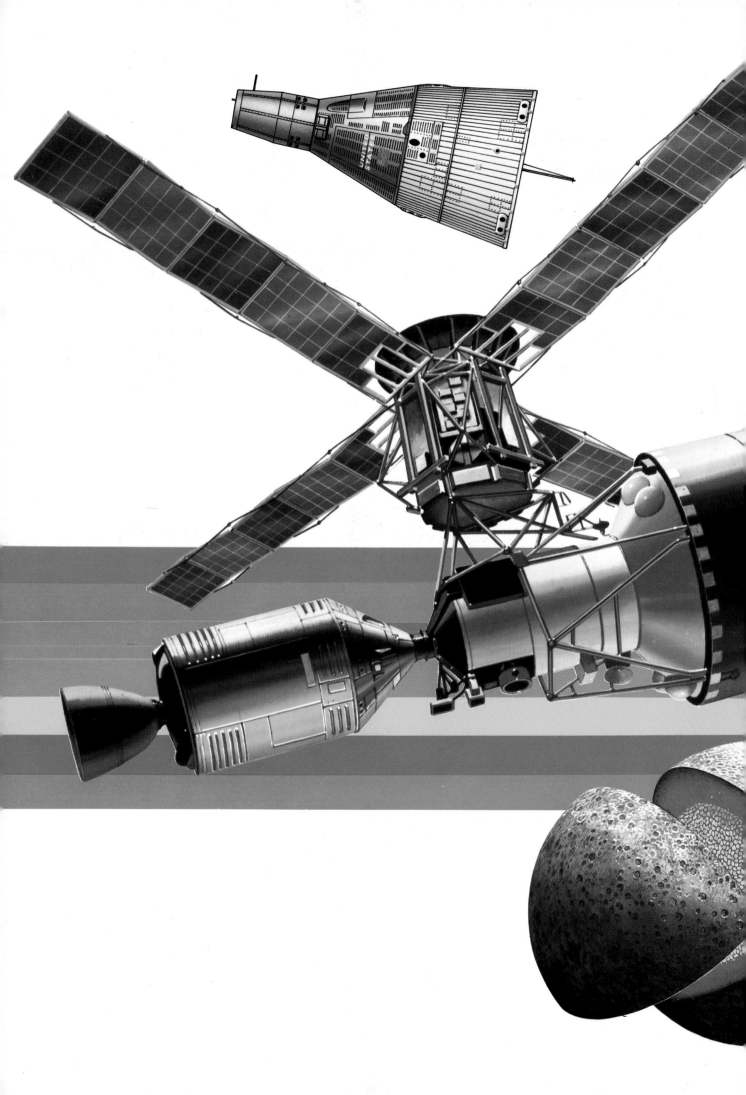